BIOFILMS

BIOFILMS

Investigative Methods & Applications

EDITED BY

Hans-Curt Flemming, Ph.D.
IWW, University of Duisburg

Ulrich Szewzyk, Ph.D.
Technical University of Berlin

Thomas Griebe, Ph.D.
IWW, University of Duisburg

CRC Press
Taylor & Francis Group
Boca Raton London New York

CRC Press is an imprint of the
Taylor & Francis Group, an **informa** business

CRC Press
Taylor & Francis Group
6000 Broken Sound Parkway NW, Suite 300
Boca Raton, FL 33487-2742

First issued in paperback 2019

ISBN-13: 978-1-56676-869-6 (hbk)
ISBN-13: 978-0-367-39842-2 (pbk)

Visit the Taylor & Francis Web site at
http://www.taylorandfrancis.com

and the CRC Press Web site at
http://www.crcpress.com

Main entry under title:
   Biofilms: Investigative Methods and Applications

A Technomic Publishing Company book
Bibliography: p.
Includes index p. 241

Library of Congress Catalog Card No. 00-102583

Table of Contents

Preface

DURING the last 10 years biofilms have become an important object of microbiological inquiry as a critical element in the preservation of quality within water systems as well as a key component of biological reactions in wastewater treatment. An understanding of biofilm development, structure and dynamics is one condition for improving water supplies and for addressing technical problems such as biofouling, corrosion and bioweathering.

Until recently the understanding, as well as management, of biofilms were impeded due to a dearth of investigative techniques. The present volume is the first to organize and examine the best currently available methods for investigating biofilms. It covers culture-based methods and emerging nondestructive techniques and shows how they can be used to characterize biofilms in a variety of manmade settings, such as sewers, wastewater plants, and drinking water distribution systems, as well as in karsts and groundwater sources. At the same time the book explains technologies for the controlled growth of biofilms and shows how biofilms can be effectively monitored and subjected to quantitative analysis. The technical information in this volume is designed to be of use to engineers and researchers and to be helpful in the generation of electronic data.

The methods presented in this book were selected under several criteria: robustness, proven utility, and applicability to the solution of a variety of biofilm-related problems. It is the editors' intent that it will serve as an ongoing reference for all specialists called on to characterize and control the unique microbial communities that constitute biofilms.

List of Contributors

B. Aßmus
GSF-Forschungszentrum für Umwelt und Gesundheit
Institut für Bodenökologie
Neuherberg Postfach 1129
85758 Oberschleissheim
Germany

Nicholas J. Ashbolt
Center for Waste and Water Technology
School of Civil and Environmental Engineering
University of New South Wales
Sydney, NSW, 2052
Australia

Peter J. Beatson
UNESCO Center for Membrane Science and Technology
School of Chemical Engineering and Industrial Chemistry
University of New South Wales
Sydney, NSW, 2052
Australia

Hans-Curt Flemming
Universität Duisburg
Fachbereich 6
Aquatische Mikrobiologie
Geibelstrasse 41
D-47057 Duisburg
Germany

Janine A. Flood
Dept. of Microbiology
Montana State University
Bozeman, MT 59717
U.S.A.

Thomas Griebe
Universität Duisburg
Fachbereich 6
Aquatische Mikrobiologie
Geibelstrasse 41
D-47057 Duisburg
Germany

A. Hartmann
GSF-Forschungszentrum für Umwelt und Gesundheit
Institut für Bodenökologie
Neuherberg Postfach 1129
85758 Oberschleissheim
Germany

Martina Hausner
Technische Universität München
Lehrstuhl für Wassergüte- und Abfallwirtschaft
Am Coulombwall
85748 Garching
Germany

Sandra Hoffmann
WFM Wasserforschung Mainz GmbH
Rheinallee 41
55118 Mainz
Germany

Harald Horn
Hydrochemie
FH Magdeburg
Brandenburgerstr. 9
D-39104 Magdeburg
Germany

Andreas Jahn
Environmental Engineering Laboratory

Äalborg University
Sohngaardsholmsvej 57
DK-9000 Aalborg
Denmark

Sibylle Kalmbach
Technische Universität Berlin
FB Ökologie der Mikroorganismen
Franklinstrasse 29
D-10587 Berlin
Germany

Martin Kuehn
Technische Universität München
Lehrstuhl für Wassergüte- und Abfallwirtschaft
Am Coulombwall
85748 Garching
Germany

John R. Lawrence
National Hydrology Research Institute
11 Innovation Boulevard
Saskatoon, Saskatchewan
S7N 3H5
Canada

Werner Manz
Technische Universität Berlin
FB Ökologie der Mikroorganismen
Franklinstrasse 29
D-10587 Berlin
Germany

E. Müller
Bayerisches Landesamt für Wasserwirtschaft
Institut für Wasserforschung
Kaulbachstr. 37
D-80539 München
Germany

Thomas R. Neu
Department of Inland Water Research Magdeburg
UFZ Centre for Environmental Research

Leipzig-Halle
39114 Magdeburg
Germany

Per Halkjaer Nielsen
Environmental Engineering Laboratory
Äalborg University
Sohngaardsholmsvej 57
DK-9000 Aalborg
Denmark

Ursula Obst
WFM Wasserforschung Mainz GmbH
Rheinallee 41
55118 Mainz
Germany

Isolde Roske
Technische Universität Dresden
Institut für Mikrobiologie
Mommsenstr. 13
01062 Dresden
Germany

Kerstin Roske
Technische Universität Dresden
Institut für Mikrobiologie
Mommsenstr. 13
D-01062 Dresden
Germany

Wolfgang Sand
Universität Hamburg
Institut für Allgemeine Botranik und Botanischer Garten
Abteilung Mikrobiologie
Ohnhorststrasse 18
D-22609 Hamburg
Germany

Gabriela Schaule
IWW Institut für Wasserforschung
Moritzstr. 26
D-45476 Mülheim
Germany

Ulrich Schindler
Technische Universität München
Lehrstuhl für Wassergüte- und Abfallwirtschaft
Am Coulombwall
85748 Garching
Germany

Michael Schloter
GSF-Forschungszentrum für Umwelt und Gesundheit
Institut für Bodenökologie
Ingolstaedter Landstr. 1
85758 Oberschleissheim
Germany

Thomas Schwartz
WFM Wasserforschung Mainz GmbH
Rheinallee 41
55118 Mainz
Germany

Klaus-Peter Seiler
GSF-Forschungszentrum für Umwelt und Gesundheit
Institut für Hydrologie
Ingolstaedter Landstr. 1
85758 Oberschleissheim
Germany

Rosemarie Spaeth
Technische Universität München
Lehrstuhl für Wassergüte- und Abfallwirtschaft
Am Coulombwall
D-85748 Garching
Germany

Ulrich Szewzyk
Technische Universität Berlin
FB Ökologie der Mikroorganismen
Franklinstrasse 29
D-10587 Berlin
Germany

Henry von Rège
Universität Hamburg

Institut für Allgemeine Botranik und Botanischer Garten
Abteilung Mikrobiologie
Ohnhorststrasse 18
D-22609 Hamburg
Germany

Peter A. Wilderer
Technische Universität München
Lehrstuhl für Wassergüte- und Abfallwirtschaft
Am Coulombwall
D-85748 Garching
Germany

Axel Wobus
Technische Universität Dresden
Institut für Mikrobiologie
Mommsenstr. 13
D-01062 Dresden
Germany

Gideon M. Wolfaardt
Dept. of Applied Microbiology
University of Saskatchewan
Saskatoon, Saskatchewan
S7N 0W0
Canada

Stefan Wuertz
Technische Universität München
Lehrstuhl für Wassergüte- und Abfallwirtschaft
Am Coulombwall
D-85748 Garching
Germany

CHAPTER 1

Steps in Biofilm Sampling and Characterization in Biofouling Cases

GABRIELA SCHAULE
THOMAS GRIEBE
HANS-CURT FLEMMING

INTRODUCTION

IF problems that are suspected of originating from biofouling arise in technical water systems, it is important to verify this assumption to design appropriate countermeasures. Because biofouling is generally a biofilm problem (Characklis et al., 1990), detection of biofilms is the analytical key to verify the diagnosis "biofouling."

The first step in a rational anti-fouling strategy frequently will be the analysis of deposits. In most cases, the deposits in technical water systems will not consist only of biomass but will contain considerable amounts of abiotic inorganic and organic material, e.g., encrustations of mineral deposits, corrosion products or organic conditioning agents. The biological analysis verifies the presence of microorganisms and their proportion of the overall composition of the deposit.

In general, biofilms contain the following main components:

- water (often > 90%)
- EPS (up to 90% of organic matter)
- cells
- entrapped particles and precipitates
- sorbed ions and polar and apolar organic molecules

In some instances, a biofilm may accumulate many inorganic particles because of the adhesive properties of the EPS. These particles can contribute more to the overall mass of the deposit than the biofilm itself. To understand the fouling process, it is important in such cases to consider that the biofilm acts as a "glue" that causes the deposition of the material in the particular place. In fouling situations, the proportion of microorganisms can be even

1

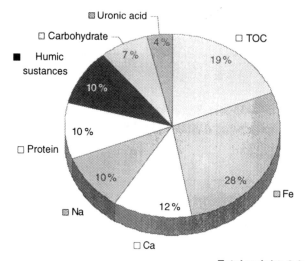

Total weight: 2.2 g m^{-2}
Total cell number: 6.4 x 10^8 cm^{-2}

Figure 1 Composition of a deposit from a reverse osmosis membrane.

smaller, and perhaps almost imperceivable within thick, crusty deposits. However, it is important to detect them, because they possibly play a crucial role in the formation of such deposits. An example is the excessive deposition of ferric oxide-hydroxide by *Gallionella* sp., a case in which very few microorganisms and, consequently, a very small amount of biomass causes a large deposition of inorganic material. Thus, biofouling analysis in technical systems usually also has to deal with abiotic components. A typical example for the result of a deposit analysis is given in Figure 1.

SAMPLING STRATEGY

The most common way to address biofouling problems is sampling of the water phase. This can usually be performed relatively easily but will provide no information about the location and the extent of deposits on surfaces. It is very important to keep in mind that microbial analysis of the water phase is not suitable to accurately locate or quantify biofilms because the contamination of the water phase occurs not continuously but randomly and does not reflect the site or extent of biofilms in a system. This emphasis on sampling of surfaces may sound trivial; however, in practice, surface sampling is often difficult.

In the first place, samples can be taken either by removing material directly from surfaces or by removing parts of the system carrying the fouling layers. This decision has to be taken at the site where the fouling problem occurs. In case the sample is taken directly, preliminary investigations must be performed to design the most suitable protocol for effective removal of the material from the surface. The sampling surface area has to be determined. If it is necessary to remove parts of a system, e.g., filter or ion exchanger material, it is mandatory to keep the samples under constant humidity. Transport should be performed cooled and mechanical shocks should be avoided.

In systems that are prone to biofouling, it has proven very useful to incorporate removable surface parts, commonly called "coupons." This facilitates the monitoring of biofilm accumulation. In drinking and process water systems, bypass monitoring systems are in use such as the biofilm monitoring device, based on ATP measurement (van der Kooji et al., 1995). Another option is the exposure of coupons under the hydrodynamic conditions of, e.g., the piping walls, known as "Robbins devices" (Ruseska et al., 1982). Exposure of surfaces in bypass systems is also used for monitoring. A suitable device is an annular reactor in which a cylinder rotates in a vessel, providing a defined shear force. The inner walls of the vessel carry specimens of materials on which biofilm accumulation is to be investigated; a prototype is called "RotoTorque" (Characklis, 1990) and has been modified to meet even more requirements in biofouling monitoring (Griebe and Flemming, 1996). In reverse osmosis membrane biofouling cases, sacrificial elements are used in a bypass, allowing access to representative surfaces (Winters et al., 1983; Ridgway et al., 1984). All these systems clearly have their merits; however, they require a good knowledge of the principle on which they are based and a good deal of experience to interpret the results. Further developments are being made, and there is still a strong demand for simple, accurate, non-destructive and real-time monitoring.

FIELD METHODS IN SITU

Some simple field methods can give first clues about biofouling. In many cases, optical inspection can give a good indication of the presence of biofilms, and in some cases, characteristic odors confirm the diagnosis. Biofilms in technical systems have usually reached the plateau phase of growth and are in most cases visible to the naked eye. In technical systems, they tend to display a slimy consistency that can be detected by wiping; thin layers can be made visible if a white tissue is used for wiping. In case of doubt about the biological origin of a deposit or layer, it is useful to take a small amount of the material and put it over the flame of a lighter until it smolders. A smell of burnt protein is characteristic for biological material.

LABORATORY METHODS

In many cases, field methods are not sufficient to define biofouling and laboratory investigations are required. A wide range of methods is available for this purpose and the selection of a particular method depends essentially on the question to be answered, e.g., is the biofilm hygienically relevant? Does it block filters or membranes? Does it contaminate ultrapure water?

Also, the substratum surface texture, chemistry and geometry, and the nature of the deposit have to be considered. A first differentiation is possible by microscopy, followed by biochemical methods. Although all methods mentioned in this paper are valid for biofilms removed from their substratum, only a few of them are applicable for direct analysis of the non-disturbed sample. These are discussed in the following paragraphs.

METHODS TO DIRECTLY EXAMINE ADHERENT MICROORGANISMS

Microscopy

The most common method for enumeration and morphological observation of microorganisms on surfaces is microscopy. This includes direct counting methods such as light microscopy, epifluorescence microscopy, scanning electron microscopy (SEM), and confocal laser scanning microscopy (CLSM). If the biofilm is thin, direct epifluorescence microscopy will be the most feasible method, using autofluorescence staining with 4,6-diamidino-2-phenylindole (DAPI) or acridine orange (AO) as DNA-specific dyes, indicating the total number and distribution of cells. Biofilms of more than 3- to 4-μm thickness usually cannot be handled with common light microscopes because material above and below the focal plane will scatter light and interfere with the direct measurement. Such biofilms can be investigated nondestructively either by computer enhanced microscopy (Walker and Keevil, 1994) or by CLSM (Lawrence et al., 1991; Caldwell et al., 1992a). The CLSM allows optical sectioning of the structure of the biofilms. With image analysis, the three-dimensional reconstruction of the undisturbed sample is possible. This is of particular interest for biocorrosion studies.

Direct enumeration methods have to take into account the possible interference of the substratum, and it may be necessary to remove the microorganisms to be able to enumerate or to characterize them. Polymer materials, such as reverse osmosis membranes, are often autofluorescent over a wide spectral range and fluorescent dyes are of no value. Therefore, removal, homogenization, dilution or concentration by membrane filtration may be required prior to enumeration.

Scanning electron microscopy (SEM) provides another option for the investigation of biofilms; however, it must be taken into account that the process of sample preparation includes complete dewatering. Considering that biofilms consist mainly of water, the picture obtained is necessarily an artifact. It still gives insight into the structure of the biofilm and can reveal the presence of different kinds of extracellular polymer substances (EPS). To avoid excessive shrinking of the sample, the EPS are stabilized with Ruthenium Red, and dewatering can be performed by using a fluorocarbon compound that allows drying by sublimation (Griebe and Flemming, 1996). A new SEM technique called "Environmental Scanning Electron Microscopy" (ESEM) overcomes this obstacle. With this method it could be demonstrated that the sample preparation process for classical SEM leads to a considerable loss of material (Little and DePalma, 1988). However, the magnification achieved by ESEM is currently much lower than with the conventional SEM.

Nucleic Acid-Specific Dyes

DNA and RNA provide large numbers of intracellular binding sites that promote marked fluorescence enhancement of many stains (fluorochromes) such as DAPI, AO, Hoechst 33258 or 33342, SYTOX green, SYBR Green, PicoGreen and propidium iodide (Roth et al., 1997). Their application is supposed to allow detection of the organisms in a deposit and to distinguish them from particles.

Vital Dyes

The vitality of the microorganisms in the biofilm can provide very important information to assess the efficacy of biocides. This can be performed by using vital dyes. They allow the recognition and localization of physiologically active bacteria in biofilms. A prerequisite is a biochemically induced reaction within the cell. If a fluorescent dye is used, the reaction can be observed with incident light using an epifluorescence microscope or a CLSM. An example is the reduction of the colorless, water-soluble dye 5-cyano-2,3-ditolyl tetrazolium salt (CTC) to an insoluble and fluorescent formazan deposit by the cellular respiration chain, which fluoresces in red after excitation (Rodriguez et al., 1992; Schaule et al., 1993). Further examples are other fluorogenic substrates such as sulfofluorescein diacetate (SFDA) or carboxyfluorescein diacetate (CFDA, Brul et al., 1997). Enzymatic activity leads to a hydrolytic cleavage of substrate and a fluorochrome that fluoresces after excitation.

Apart from the determination of intracellular enzymatic activity and redox potential, the integrity of the cell membrane is generally considered as an indicator of the viability of bacterial cells. Probes that detect the membrane integrity include positively charged fluorochromes such as RH-795, carbocyanine derivatives and Rhodamine 123 (Morgan et al., 1993), or the exclusion of negatively charged bisoxonols by active and intact bacteria (Jepras et al., 1995; López-Amorós et al., 1996; Mason et al., 1994; 1995). The most rapid and commonly used test kit for the assessment of viability (Live/Dead Bac-Light®, Molecular Probes) is based on the staining with the membrane potential sensitive dye propidium iodide and the nucleic acid stain SYTO 9 (Roth et al., 1997). Lawrence et al. (1996) reported problems associated with the use of this test kit when applied to biofilms. The green nucleic acid stain SYTO 9 often failed to penetrate microcolonies of living cells that were embedded in EPS, which led to an underestimation of the number of viable cells.

Antibiotics, such as nalidixic or pipemidic acid, prevent cell division and allow the detection of living bacteria by their elongation after incubation with substrate and staining with a fluorochrome (Kogure et al., 1979; Yu et al., 1993).

Redox Activity

Dehydrogenase activity can be determined by the use of tetrazolium dyes. Examples are triphenyl tetrazolium chloride (TTC, Ryssov-Nielsen, 1975), 2-p-(iodophenyl)-3-(p-nitrophenyl)-5-tetrazolium chloride (INT, Chung and Neethling, 1989) and CTC (Rodriguez et al., 1992; Schaule et al., 1993). The usual procedure is to incubate the sample with the reduction dyes, extract the formazans after the reaction time and determine the overall fluorescence intensity. In the presence of elevated levels of non-biological material or in thick biofilms, INT and TTC are less suitable than CTC to indicate microbial redox activity because they form non-fluorescent formazans in contrast to CTC, with the consequence that the crystals cannot be visualized by incident light and, thus, cannot be detected directly in biofilms on surfaces. In strongly reduced samples, the redox potential can be as low as -200 mV with the consequence that INT and TTC are reduced abiotically to formazan, whereas CTC is reduced only from -220 mV (Seidler, 1991). Another useful redox dye is resorufin, which has been used as an indicator of cytoplasmic reducing potential in yeast cells (Caldwell et al., 1992b).

Biofilm Thickness

The thickness of a biofilm or a fouling layer can be of interest, e.g., in cases of membrane fouling in RO processes (Griebe and Flemming, 1996). If the layer is sufficiently transparent, it is possible to assess the thickness by microscopic methods (Bakke and Olson, 1986), but in most cases, the layer will be opaque and of gray, reddish or brownish color. In such cases, thickness determination by cryosectioning can be an interesting alternative (Yu et al., 1993). The steps are schematically depicted in Figure 2.

Determination of Biofilm Thickness

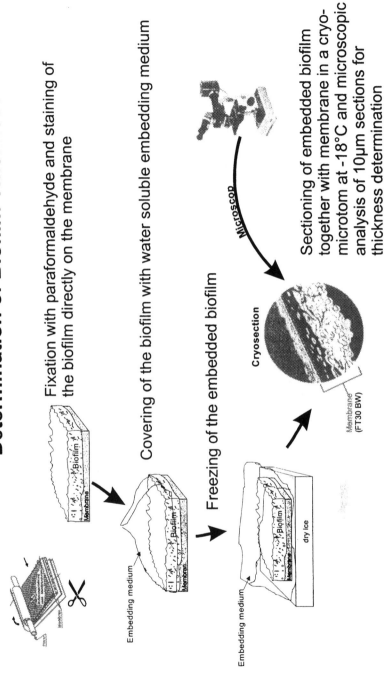

Fixation with paraformaldehyde and staining of the biofilm directly on the membrane

Covering of the biofilm with water soluble embedding medium

Freezing of the embedded biofilm

Sectioning of embedded biofilm together with membrane in a cryo-microtom at -18°C and microscopic analysis of 10µm sections for thickness determination

Embedding medium

Embedding medium

dry Ice

Microscop

Cryosection

Membrane (FT30 BW)

Figure 2 Steps in cryosectioning for the determination of biofilm thickness and structure in biofouling analysis on reverse osmosis membranes.

7

In special cases, mostly for research purposes, it can be of importance to elucidate structure and composition of biofilms in further detail. Detailed information about microcolony cluster size, porosity, voids and cell density can be acquired by using specific stains and staining techniques in combination with cryosectioning (Figure 3).

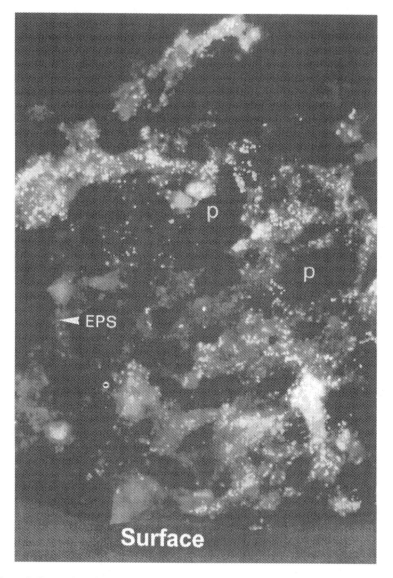

Figure 3 Cryosection of an environmental biofilm. A thin cross-section (0.5 μm thickness) enables the differentiation of single bacteria (AO stained) in clusters, pores (labeled P) and EPS (arrow) stained with FITC.

FTIR-ATR Spectroscopy for Characterization of Biofilm Composition

FTIR spectroscopy has been used in biofilm investigation for quite some time (Nivens et al., 1985; Schmitt and Flemming, 1996) and is particularly useful if biotic and abiotic components occurring in a deposit are to be differentiated. Much practical experience was gained by the investigation of fouled membranes; in many cases biological and other kinds of fouling usually coincide and have to be differentiated (Flemming et al., 1996). To do so, FTIR spectroscopy is applied in the attenuated total reflection (ATR) mode. This technique allows the analysis of smooth surfaces directly. Practically no sample preparation is required. The acquired absorption spectra reveal the chemical nature of the deposit. Biological material is detected by the occurrence of the typical amide I and II bands that originate from proteins (Schmitt and Flemming, 1998).

METHODS TO REMOVE MICROORGANISMS FROM SURFACES

Mechanical Force

Mechanical removal of biofouling layers includes scraping with various tools such as scrapers or flat knives, or abrasive removal using glass beads or other particulate material. To remove and homogenize thick layers, often a combination of different scraping methods followed by ultrasonic treatment is feasible.

Biofilm detection by direct swab collection is one of the few standardized methods to remove thin biofilms. A defined area of the wet surface is swabbed with a polypropylene swab. The collected material is transferred into a bottle of sterile water and analyzed. The method is validated and documented in the *Sematech Provisional Test Method for Determining the Surface Associated Biofilms of UPW Distribution Systems 92010958B-STD* (1992). This method is applicable for smooth system surfaces and for parts that can be removed from the system. It is not feasible for pipes with small diameter, or rough inner surface or if areas a long distance from the access point have to be sampled. In such cases, other methods such as the use of glass beads with a smooth surface and approximately 0.5 mm in diameter can be used for abrasion. This method was first introduced for ultrapure water systems by Schaule and Flemming (1997).

An Example: Biofouling in a Ultrapure Water (UPW) System

The irregular occurrence of high numbers of colony-forming units in the water phase indicates the presence of a biofilm contaminating a UPW system. To locate the contamination source, divide the system into different sections from which water samples are taken. These samples are analyzed microscopically for the presence of flocs, which are indicative of biofilms. Such flocs are not detected by cultivation methods with subsequent plate counting. With this strategy, the most contaminated section of the system can be identified. Then, parts of the section situated in front of that area are removed (e.g., pipes), and the fouling layer is completely removed by using the following protocol:

(1) One end of the pipe is closed aseptically. A defined volume of sterile water and sterile glass beads are added.

(2) The other end is closed aseptically. The pipe is carefully shaken for several minutes and turned around the long axis to allow contact of the glass beads to the entire inner surface.

(3) One end is opened, and the glass beads and the water are collected in a sterile bottle.

(4) The supernatant is separated after approximately 1 min and analyzed for any interesting parameter.

Ultrasonic Treatment

Ultrasonic treatment is a widely used and reliable method to remove and disaggregate biofilms. To avoid cell lysis, usually a sonication time of 1–3 min is chosen and with commercially available sonication baths, a frequency of 40–50 kHz, obtains the best results in terms of removal and survival of cells. It should also be taken into account that the position of the sample relative to the sonicator can change the effect significantly (Zips et al., 1990). This is one of the results of a round robin test carried out by a European task group for evaluation of the removal of biofilms. The results of the round robin tests showed that 3 min of sonication was in most cases sufficient for effective biofilm removal. Nevertheless it should be noted that the yield was strongly dependent on frequency, energy, distance to the sample and geometry of the sonication device.

LABORATORY METHODS TO EXAMINE REMOVED BIOFILMS

METHODS REQUIRING CULTIVATION

Classical microbiological identification methods are based on cultivation of the organisms. Because all media are selective to a lesser or greater extent and not all bacteria or other microrganisms are culturable, viable counts rarely represent all microorganisms present in a sample. Therefore, it is useful to determine the total cell number by the methods described earlier. This number allows an estimate of the ratio of culturable cells.

COLONY-FORMING UNITS (CFU) AND MOST PROBABLE NUMBER (MPN) METHODS

CFU plate counting is the most commonly used method to monitor water quality. There are standardized methods documented for drinking water, ultrapure water, water for injection, and wastewater (e.g., national legal regulations and pharmacopoeae). In biofouling cases, it has proven useful to expand the range of media and incubation parameters to apply the most sensitive method. For better comparability it is generally recommended to use R2A medium (Reasoner and Geldreich, 1985) and an incubation temperature of 20°C and 28°C as defined in the ASTM. The results should be documented as CFU at 20°C after 7 days to include organisms that are not detected after shorter incubation times.

The application of cultivation methods to homogenized biofilms can hold certain disadvantages. If the proportion of non-biological components of the deposit is high, the membrane filtration as a method for concentrating cells is of limited value. In such cases, the most probable number (MPN) method is useful. This is a standard procedure of classical microbiology. When assessing, e.g., the presence of *E. coli,* the number is not related to a volume of 100 mL but to a defined surface area or a defined quantity of the deposit.

In a case study, various methods for the determination of viable and vital bacteria from a reverse osmosis (RO) membrane biofilm were compared (Figure 4). This example clearly demonstrates the differences between plate counting techniques and microscopic techniques for the detection of vital bacteria.

IDENTIFICATION

IDENTIFICATION BY ANALYSIS OF PHOSPHOLIPID FATTY ACIDS

The analysis of the phospholipid fatty acid (PLFA) profile has been developed for the characterization of microbial populations in soil and sediment samples. It is based on the composition of the phospholipids that are present in the membranes of all living cells. They can be extracted and quantified with excellent sensitivity (Tunlid and White, 1990a). This method allows the assessment of the quantity of biomass at a given site. Phospholipids belong to the first compounds to be degraded when cells are decaying (within hours); therefore, the content of PLFA indicates the amount of living biomass. Because the composition of fatty acids is different in different organisms, the PLFA profile can be used to analyze the composition of the population (White, 1993; Zelles and Bai, 1993) and is used as a taxonomic marker (LeChevallier, 1977; Vestal and White, 1989).

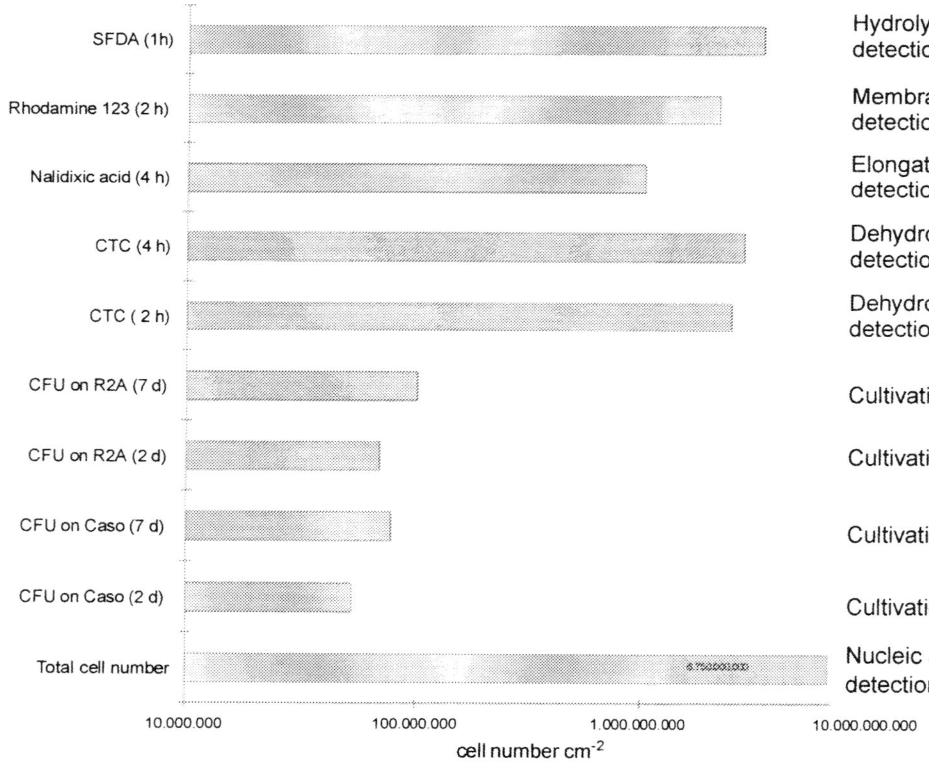

Parameter and detection	Percent of total cell number
Hydrolytic (esterase) activity (microscopic detection)	50
Membrane permeability (microscopic detection)	32
Elongation (gyrase activity) (microscopic detection)	15
Dehydrogenase activity (microscopic detection)	41
Dehydrogenase activity (microscopic detection)	37
Cultivation on low nutrient medium	1,5
Cultivation on low nutrient medium	1,0
Cultivation on nutrient rich medium	1,2
Cultivation on nutrient rich medium	0,8
Nucleic acid marker (microscopic detection)	100

Figure 4 Comparison of data obtained from a RO-membrane biofilm sample by microscopic and cultivation methods for the determination of total and viable bacteria. The preparation procedure, incubation time, and cell numbers are shown.

An advantage of the method is that the samples can be analyzed directly and without prior treatment. However, it is mainly suitable for relatively high concentrations of biomass as encountered in soil and sediment samples, whereas populations of thin biofilms will be difficult to analyze with this method. Comparison with other methods such as the content of ATP, DNA and protein revealed a good correlation (White, 1988). The qualitative and quantitative determination of PLFA is performed by gas chromatography or gas chromatography/mass spectrometry. The method provides the following information:

- quantity of living biomass (Tunlid and White, 1990b)
- gross characterization of population composition (diversity of population, dominance of certain groups) (Tunlid and White, 1990b)
- nutrient situation (Kieft et al., 1994)
- presence of toxic or inhibiting substances (Napolitano et al., 1994)

FT-IR SPECTROSCOPY

A more rapid method to identify isolates is given by FTIR spectroscopy. Most structural and functional groups of different bacteria are identical and give the same signals. Thus, spectra of bacteria look very much alike. However, the quantity and distribution of the different groups vary significantly among microbial strains, and this is the basis for all approaches to the characterization of bacteria by their IR spectra. The advantages of the modern FTIR spectrometers, the software and elaborated mathematical and statistical algorithms for analyzing IR-spectra allows the detection of the differences well enough to distinguish different bacterial strains, and identify them, if there are references available (Schmitt et al., 1995). A considerable library of reference spectra for drinking water bacteria has been established. The occurrence of bacterial storage products as polyhydroxyalkanoates, such as polyhydroxybutyrate (PHB) or other biopolymers can be detected in FTIR spectra as well. This allows the observation of physiological reactions of microorganisms, because these substances are formed under stress and unbalanced nutritional conditions.

CHEMICAL AND BIOCHEMICAL ANALYSIS

WATER CONTENT

The determination of water occurs by measurement of the sample before and after drying at 105°C to a constant weight. In most cases, it will be necessary to remove them from the surface in question to determine the weight. The exact quantification of the water content can be difficult in cases when thin biofilms are exposed to the atmosphere and the water evaporates quickly. In such cases, it may be useful to remove the fouling layer under 100% humidity.

TOTAL ORGANIC CARBON (TOC)

The total organic carbon (TOC) content gives an idea about the proportion of organic matter in a sample and can be interpreted as biological material, e.g., in corrosion products in which no other organic material is to be expected.

The chemical and biochemical analysis is one approach to measure the biomass of biofilms. A number of intra- and extracellular components, such as total organic carbon (TOC), protein, lipopolysaccharide, carbohydrate or muramic acid content, have been successfully used to quantify universally distributed biofilm components. Some parameters indicative for the presence of biofilms are summarized in Table 1.

POLYSACCHARIDES AND PROTEINS

Calcofluor White (Wood, 1980), Primulin (Caron, 1983) and conjugated lectins (Wolfaardt et al., 1995; Caldwell et al., 1992a) bind to polysaccharides. Lectins that bind to specific combinations of sugar molecules have been used for the identification and localization of cell types (e.g., gram positive or negative) or components in the matrix of EPS (Lawrence et al., 1994). A serious

TABLE 1. Examples for Biofilm Parameters.

Parameter	Detection Method (Reference)
Water content	24 h, 105°C
Organic carbon	TOC, COD, incineration loss
Protein	(Lowry et al., 1951; Bradford, 1976)
Carbohydrates	(Dubois et al., 1956; Raunkjær et al., 1994)
Uronic acids	(Blumenkranz and Asboe-Hansen, 1973)
DNA	(Palmgren and Nielsen, 1996)
Lipids	(Geesey and White, 1990)
Muramic acid	(Geesey and White, 1990)
Polyhydroxybutyrate	(Geesey and White, 1990)
Total cell number	(Hobbie et al., 1977)
Colony-forming units	various standard methods
ATP	(Karl, 1993)
Hydrolase activity	(Obst and Holzapfel-Pschorn, 1990)
Respiratory activity	(Schaule et al., 1993)
Indolacetic acid production	(Bric et al., 1991)
Catalase activity	(Line, 1983)

pitfall of the uncritical use of lectins such as concanavalin A (Con A) and wheat germ agglutinin (WGA), which are conjugated with fluorescence dyes such as fluorescein isothiocyanate (FITC), is their non-specific binding potential to proteins (Babuik and Paul, 1969; Hara and Tanou, 1989). This leads sometimes to an accumulation of lectins in the EPS matrix of biofilms. In recent publications, lectins were coupled to different fluorochromes (TRITC, AMCA-S, Oregon Green, tetramethylrhodamine and Texas Red) (all available from Molecular Probes; www.probes.com). This allows the application of more than one lectin at a time to selectively label and map different sugar molecules because the signals can be differentiated by their specific fluorescence emissions.

The determination of the content of protein (Lowry et al., 1951) and carbohydrates (Dreywood, 1946; Dubois et al., 1956) can also be used to roughly assess the content of biomass. These parameters are of particular interest if the samples contain a high proportion of non-biological material, e.g., sediments, soils and corrosion products. However, interference with humic substances can lead to erroneous results and may require adequate modifications of the methodology as done for the above cited samples by Raunkjær et al. (1994) and Frølund et al. (1995).

ADENYLATE CONTENT

The content of ATP or the overall adenylate content can be used to quantify the biological activity in a biofilm. To do so, the adenylates are extracted from the sample and subsequently quantified by bioluminescence (luciferin/luciferase system) by HPLC chromatography (Karl, 1993). The ATP content depends on the state of activity of the organisms in the biomass. It must be taken into account that transport and storage of the sample is of strong influence to the physiological activity of the microorganisms (Griebe et al., 1997). Thus, sampling and handling details can influence the data strongly. Nevertheless, the determination of the ATP content belongs to the common methods in biofilm research (Geesey and White, 1990).

HYDROLYTIC ACTIVITY

Independent of the redox potential, microbial activity can be determined by methylumbelliferone or methylcoumarinylamide substrates, indicating the activity of hydrolytic enzymes such as glucosidases, lipases and aminopeptidases. A correlation, however, of enzymatic activity and the number of sessile microorganisms or the overall biomass is not yet possible. A considerable proportion of the hydrolytic enzymes is located in the EPS matrix (Frølund et al., 1995), and it is not known for how long they are active after their mother cells may have died.

GENETIC METHODS

POLYMERASE CHAIN REACTION (PCR)

Based on the polymerase chain reaction (PCR), various analytical techniques can be used to identify biofilm organisms without prior cultivation. First, DNA fragments have to be extracted from the sample. It may be suitable to remove the organisms, e.g., by application of ultrasonic energy, after which they are concentrated by membrane filtration and subsequently lysed (Holben et al., 1988). It is also possible to lyse the cells directly in the sample and extract the DNA by ultracentrifugation (Steffan et al., 1988; Ogram et al., 1988). The use of polyvinylpolypyrrolidone is recommended to remove humic substances if these are present in the sample and can interfere with the PCR reaction. After amplification of the extracted rDNA a mixture of amplification products is obtained, which has to be separated by molecular biological techniques and cloned subsequently. Restriction analysis of the cloned rDNA fragments is performed, and the pattern of the fragment analysis can be compared with that of the cultivable organisms of the population. This way it is possible to obtain genetic fingerprints of bacteria, protozoa, fungi, metazoa and algae possibly present in the sample. An efficient analysis of the microorganisms by means of PCR fingerprinting can be supported by using DNA sequencers, capillary electrophoresis, image processing and databases with reference patterns for the identification of unknown isolates (Tichy et al., 1996).

GENE PROBES

With rRNA-directed and fluorescence-labeled oligo nucleotide probes it is possible to detect and localize microorganisms that do not necessarily have to be culturable and to locate them in the phylogenetic tree (Ogram and Sayler, 1988; Wagner and Amann, 1996). More than one oligonucleotide can be applied to label more than one kind of organism. This method provides not only information about the species actually present in the undisturbed sample but also reveals their position in relation to each other, using the potential of the CLSM to demonstrate three-dimensional structures. This makes it particularly suitable to investigate the functional architecture of a biofilm. In addition, the intensity of the signal depends on the ribosome content and, thus, reflects the physiological status. To apply the gene probes, the cells must become permeable. The procedure required for that purpose will kill them, and it cannot be completely excluded that the architecture of the biofilm is untouched by the procedure. Nevertheless, this method belongs to the most important progress in biofilm research and is, therefore, separately discussed in Chapters 7 and 8 of this volume.

IMMUNOLOGICAL METHODS

Detection and identification of biofilm population members can also be performed by immunological coupling reactions between fluorescent labeled antibodies that are specific to certain organisms (Gaylarde and Cook, 1990; Griffin and Antloga, 1985). This method requires the cultivation of the target organisms and the subsequent production of antibodies and is suitable to detect them in a thin biofilm (Chapter 9 of this volume). In multilayered biofilms, the EPS matrix acts as a diffusion barrier for large molecules, and the antibodies will take some time to reach the antigenic sites of the target organisms. The advantage in the use of mono- and polyclonal antibodies is that the cells are not killed by this method, and, thus, processes in the biofilm still can be observed. Also, the combination with other non-destructive fluorescence staining methods is possible, e.g., with CTC.

CONCLUSIONS AND OUTLOOK

The investigation of deposits on surfaces, in particular of biofilms, represents a challenge for analytical chemistry and microbiology. Although it is possible to detect very low concentrations of chemicals or microorganisms in a given volume of water or gas, the options for the detection of deposits is still somewhat dissatisfying. At this time, it is still necessary to remove the material from the surface and investigate it subsequently. Some of the methods that apply specifically to biological material have been presented in this chapter. All these methods are more or less time consuming and yield results usually only hours or even days after sampling—this is the actual state of the art. Although the analysis of surfaces in the scale of a few Angströms or nanometers is very well developed, the analysis of surfaces in the range of square centimeters or meters still depends on scratching material from the surfaces and subsequent analysis in the laboratory.

New approaches to analyze biofilms and other deposits should provide qualitative and quantitative information on line, in situ, in real time and non-destructively. Two strategies seem to be feasible to fulfill these demands:

(1) Sensors that are installed at representative sites in a system, e.g., in a heat cooling water cycle, at a ship bottom or in a pipe line. A suitable approach is the use of fiber-optic sensors or differential turbidity measurement (Flemming et al., 1996).

(2) Probe heads that allow the scanning of surfaces in question, revealing information about contamination and deposits. This might be performed by the use of FTIR-ATR spectroscopy probe heads.

Such approaches will open a completely new field of surface analysis and in particular, to detection and monitoring of biofouling.

REFERENCES

Babuik, L.A. and E.A. Paul. 1969. The use of fluoresceinisothiocyanate in the determination of the bacterial biomass of grassland soil. *Can. J. Microbiol.* 16, 57–62.

Bakke, R. and P.Q. Olson. 1986. Biofilm thickness measurements by light microscopy. *J. Microb. Methods* 5, 93–98.

Blumenkranz, N. and G. Asboe-Hansen. 1973. New method for quantitative determination of uronic acids. *Analyt. Biochem.* 54, 484–489.

Bradford, M.M. 1976. A rapid and sensitive method for the quantification of microgram quantities of protein utilizing the principle of protein dye binding. *Analyt. Biochem.* 72, 248–254.

Bric, J.M., R. Rostock and S.E. Silverstone. 1991. Rapid in-situ assay for indolacetic acid production by bacteria immobilized on a nitrocellulose membrane. *Appl. Envir. Microbiol.* 57, 535–538.

Brul, S., J. Nussbaum and S.K. Dielbandhoesing. 1997. Fluorescent probes for wall porosity and membrane integrity in filamentous fungi. *J. Microbiol. Methods* 28, 169–178.

Caldwell, D.E., D.R. Corber and J.R. Lawrence. 1992a. Imaging of bacterial cells by fluorescence exclusion using scanning confocal laser microscopy. *J. Microb. Methods* 15, 249–261.

Caldwell, D.E., D.R. Corber and J.R. Lawrence. 1992b. Confocal laser microscopy and digital image analysis in microbial ecology. *Adv. Microb. Ecol.* 12, 1–67.

Caron, D.A. 1983. Technique for enumeration of heterotrophic and phototrophic nanoplankton, using epifluorescence microscopy, and comparison with other procedures. *Appl. Environ. Microbiol.* 46, 491–498.

Characklis, W.G., M.H. Turakhia and N. Zelver. 1990. Transport and interfacial transfer phenomena. In: W.G. Characklis and K.C. Marshall (eds.): *Biofilms.* John Wiley, New York, 265–340.

Chung, Y.-C. and J.B. Neethling. 1989. Microbial activity measurements for anaerobic sludge digestion. *J. Water Poll. Contr. Fed.* 61, 343–349.

Dreywood, R. 1946. Qualitative test for carbohydrate material. *Ind. Eng. Chem.* 18, 499.

Dubois, M.J., K.A. Gilles, J.K. Hamilton, P.A. Reber and F. Smith. 1956. Colorimetric method for determination of sugars and related substances. *Anal. Chem.* 28, 350–356.

Flemming, H.-C., J. Schmitt and K.C. Marshall. 1996. Sorption properties of biofilms. In: W. Calmano and U. Förstner (eds.): *Environmental behaviour of sediments.* Lewis Publishers, Chelsea, Michigan, 115–157.

Frølund, B., T. Griebe and P.H. Nielsen. 1995. Enzymatic activity in the activated-sludge floc matrix. *Appl. Microbiol. Biotechnol.* 43, 755–761.

Gaylarde, C. and P. Cook. 1990. New rapid methods for the identification of sulphate reducing bacteria. *Int. Biodeterior.* 26, 337–345.

Griebe, T. and H.-C. Flemming (1996): Vermeidung von Bioziden in Wasseraufbereitungs-Systemen durch Nährstoffentnahme. *Vom Wasser* 86, 217–230.

Griebe, T., G. Schaule and S. Wuertz. 1997. Determination of microbial respiratory and redox activity in activated sludge. *J. Ind. Microbiol. Biotechnol.* 19, 118–122.

Griffin, W.M. and K.M. Antloga. 1985. Development of culture media, a sporulation procedure, and an indirect immunofluorescent antibody technique for sulfate-reducing bacteria. *Dev. Ind. Microbiol.* 26, 611–626.

Hara, S. and E. Tanou. 1989. Simultaneous staining with three fluorescent dyes of minute plankters on an agarose gel filter. *Deep-Sea Res.* 36, 1777–1784.

Hobbie, J.E., R.J. Daley and S. Jasper. 1977. Use of nuclepore filters for counting bacteria by fluorescence microscopy. *Appl. Environ. Microbiol.* 33, 1225–1228.

Holben, W.E., J.K. Jansson, B.K. Chelm and J.M. Tiedje. 1988. DNA probe method for the detection of specific microorganisms in the soil community. *Appl. Environ. Microbiol.* 54, 703–711.

Jepras, R.I., J. Carter, S.C. Pearson, F.E. Paul and M.J. Wilkinson. 1995. Development of a robust flow cytometric assay for determining numbers of viable bacteria. *Appl. Environ. Microbiol.* 61, 2696–2701.

Karl, D.M. 1993. Total microbial biomass estimation derived from the measurement of particulate adenosine- 5′-triphosphate. In: P.F. Kemp, B.F. Sherr, E.B. Sherr and J.J. Cole (eds.): *Handbook of methods in aquatic microbial ecology.* Lewis Publishers, Boca Raton, Fla., 359–368.

Kieft, T.L., D.B. Ringelberg and D.C. White. 1994. Changes in ester-linked phospholipid fatty acid profiles of subsurface bacteria during starvation and desiccation in a porous medium. *Appl. Envir. Microbiol.* 60, 3292–3299.

Kogure, K., U. Simidu and N. Taga. 1979. A tentative direct microscopic method for counting living marine bacteria. *Can. J. Microbiol.* 25, 415–420.

Lawrence, J.R., G.M. Wolfaardt and D.R. Korber. 1994. Monitoring diffusion in biofilm matrices using confocal laser microscopy. *Appl. Environ. Microbiol.* 60, 1166–1173.

Lawrence. J.R., D.R. Korber, G.M. Wolfaardt and D.E. Caldwell. 1996. Analytical imaging and microscopy techniques. In: C.J. Hurst, G.R. Knudsen, M.J. McInerney, L.D. Stetzenbach and M.V. Walter (eds.): *Manual of environmental microbiology.* ASM Press, Washington, D.C., 29–52.

Lawrence, J.R., D.R. Korber, B.D. Holye, J.W. Costerton and D.E. Caldwell. 1991. Optical sectioning of microbial biofilms. *J. Bact.* 173, 6558–6567.

LeChevalier, M.P. 1977. Lipids in bacterial taxonomy—a taxonomists view. *Crit. Rev. Microbiol.* 7, 109.

Line, M.A. 1983. Catalase activity as an indicator of microbial colonization of wood. In: T.A. Oxley and S. Barry (eds.): *Biodeterioration 5.* John Wiley, New York, 38–43.

Little, B.J. and J.R. DePalma. 1988. Marine biofouling. *Treat. Mat. Sci. Technol.* 28, 89–119.

López-Amorós, R., D.J. Mason and D. Lloyd. 1996. Use of two oxonols and CTC to monitor starvation of *E. coli* in seawater by flow cytometry. *J. Microbiol. Methods* 61, 2521–2526.

Lowry, O.J., N.J. Rosebronkh, A.L. Farr and R.J. Randall. 1951. Protein measurement with folin phenol reagent. *J. Biol. Chem.* 193, 265–275.

Mason, D.J., R. Allman, J.M. Stark and D. Lloyd. 1994. Rapid estimation of bacterial antibiotic susceptibility with flow cytometry. *J. Microsc.* 176, 8–16.

Mason, D.J., R. López-Amorós, R. Allman, J.M. Stark and D. Lloyd. 1995. The ability of membrane potential dyes and calcofluor white to distinguish between viable and non-viable bacteria. *J. Appl. Bacteriol.* 78, 309–315.

Morgan, J.A.W., G. Rhodes and R.W. Pickup. 1993. Survival of nonculturable *Aeromonas salmonicida* in lake water. *Appl. Environ. Microbiol.* 59, 874–880.

Napolitano, G.E., W.R. Hill, J.B. Guckert, A.J. Stewart, S.C. Nold and D.C. White. 1994. Changes in periphyton fatty acid composition in chlorine-polluted streams. *J. North Am. Benthol. Soc.* 13, 237–249.

Obst, U. and A. Holzapfel-Pschorn. 1988. *Enzymatische tests für die wasseranalytic.* R. Oldenbourg Verlag, München, 136.

Ogram, A.V. and G. Sayler. 1988. The use of gene probes in the rapid analysis of natural microbial communities. *J. Ind. Microbiol.* 3, 218–292.

Ogram, A., G.S. Sayler and T. Barkay. 1988. DNA extraction and purification from sediments. *J. Microbiol. Methods* 7, 57–66.

Palmgren, R. and P.H. Nielsen. 1996. Accumulation of DNA in the exopolymeric matrix of activated sludge and bacterial cultures. *Water Sci. Technol.* 34, 5–6, 233–240.

Raunkjær, K., T. Hvitved-Jacobsen and P.H. Nielsen. 1994. Measurement of pools of protein, carbohydrate and lipid in domestic wastewater. *Water Res.* 28, 251–262.

Reasoner, D.J. and E.E. Geldreich. 1985. A new medium for the enumeration and subculture of bacteria from potable water. *Appl. Environ. Microbiol.* 49, 1–7.

Ridgway, H.F., M.G. Rigby and D.G. Argo. 1984. Biological fouling of reverse osmosis membranes: the mechanism of bacterial adhesion. Proc. Water Reuse Symp. II, "The Future of Water Reuse," San Diego, 1314–1350.

Rodriguez, G.G., D. Phipps, K. Ishiguro and H.F. Ridgway. 1992. Use of a fluorescent redox probe for direct visualization of actively respiring bacteria. *Appl. Environ. Microbiol.* 58, 1801–1808.

Roth, B.L., M. Pott, S.T. Yue and P.J. Millard. 1997. Bacterial viability and antibiotic susceptibility testing with SYTOX Green nucleic acid stain. *Appl. Environ. Microbiol.* 63, 2421–2431.

Ruseska I., J. Robbins, E.S. Lashen and J.W. Costerton. 1982. Biocide testing against corrosion-causing oilfield bacteria helps control plugging. *Oil Gas J.* 253–264.

Ryssov-Nielsen, H. 1975. Measurement of the inhibition of respiration in activated sludge by a modified determination of the TTC-dehydrogenase activity. *Water Res.* 9, 1179–1185.

Schaule, G. and H.-C. Flemming. 1997. Pathogenic microorganisms in water system biofilms. *Ultrapure Water,* April 1997, 21–28.

Schaule, G., H.-C. Flemming and H.F. Ridgway. 1993. The use of CTC (5-cyano-2,3-ditolyl tetrazolium chloride) in the quantification of respiratory active bacteria in biofilms. *Appl. Environ. Microb.* 59, 3850–3857.

Schmitt, J. and H.-C. Flemming. 1996. FTIR spectroscopy. In: E. Heitz, W. Sand and H.-C. Flemming (eds.): *Microbial deterioration of materials.* Springer, Heidelberg, 143–157.

Schmitt, J. and H.-C. Flemming. 1998. FTIR-spectroscopy in microbial and material analysis. *Int. Biodet. Biodegr.* 41, 1–11.

Schmitt, J., S. Krietemeyer, G. van den Bosche and H.-C. Flemming. 1995. Biofilme in der Trinkwasseraufbereitung. *DVGW 110, Schriftenreihe* 259–279.

Seidler, E. 1991. *The tetrazolium-formazan system: design and histochemistry.* G. Fischer Verlag, Stuttgart, New York, 86.

Steffan, R.J., J. Goksoyr, K.B. Asim and R.M. Atlas. 1988. Recovery of DNA from soil and sediments. *Appl. Environ. Microbiol.* 54, 2908–2915.

Tichy, H.-V., P. Wiesner and R. Simon. (1996): PCR-Fingerprint-Verfahren zur Analyse von Mikroorganismen-Populationen. In: H. Lemmer, T. Griebe and H.-C. Flemming (eds.): Ökologie der Abwasserorganismen. Springer Verlag, New York/Berlin, 111–122.

Tunlid, A. and D.C. White. 1990a. Use of lipid biomarkers in environmental samples. In: A. Fox, S.L. Morgan, L. Larsson and G. Odham (eds.): *Analytical microbiology methods.* Plenum Press, New York, London, 259–274.

Tunlid, A. and D.C. White. 1990b. Use of lipid biomarkers in environmental samples. In: A. Fox, S.L. Morgan, L. Larsson and G. Odham (eds.): *Analytical microbiology methods.* Plenum Press, New York, London, 259–274.

van der Kooij, D., H.R. Veenendaal, C. Baars-Lorist, D.W. van der Klift and Y.D. Drost. 1995. Biofilm formation on surfaces of glas and teflon exposed to treated water. *Water Res.* 29, 1655–1662.

Vestal, J.R. and D.C. White. 1989. Lipid analysis in microbial ecology. *BioScience* 39, 535–541.

Walker, J.T. and C.W. Keevil. 1994. Study of microbial biofilms using light microscope techniques. *Int. Biodet. Biodegr.* 11, 223–236.

White, D.C. 1993. In situ measurement of microbial biomass, community structure and nutritional status. *Phil. Trans. R. Soc. Lond.* A 344, 59–67.

White, D.C. 1986. Environmental effects testing with quantitative microbial analysis: chemical signatures correlated with in situ biofilm analysis by FT/IR. *Toxicity Assessment* 1, 315–338.

Winters, H., I.R. Isquith, W.A. Arthur and A. Mindler. 1983. Control of biological fouling in seawater reverse osmosis. *Desalination* 47, 233–238.

Wolfaardt, G.M., J.R. Lawrence, R.D. Robarts and D.E. Caldwell. 1995. Bioaccumulation of the herbicide diclofop in extracellular polymers and its utilization by a biofilm community during starvation. *Appl. Environ. Microbiol.* 61, 152–158.

Wood, P.J. 1980. Specificity in the interaction of direct dyes with polysaccharides. *Carb. Res.* 85, 271–287.

Yu, F.P., B.H. Pyle and G.A. McFeters. 1993. A direct viable count method for the enumeration of attached bacteria and assessment of biofilm disinfection. *J. Microb. Methods* 17, 167–180.

Zelles, L. and Q.Y. Bai. 1993. Fatty acid patterns of phospholipid and lipopolysaccharides in environmental samples. *Chemosphere* 28, 391–411.

Zips, A., G. Schaule and H.-C. Flemming. 1990. Ultrasound as a means of detaching biofilms. *Biofouling* 2, 323–333.

CHAPTER 2

Rotating Annular Reactors
for Controlled Growth of Biofilms

THOMAS GRIEBE
HANS-CURT FLEMMING

INTRODUCTION

INVESTIGATIONS of resistance, growth, accumulation, interactive and cooperative activity of functional consortia of biofilm microorganisms require test systems or laboratory devices. The choice of an appropriate system arises from difficulties to simulate, control and analyze the nature and the properties of biofilms in laboratory experiments that reflect environmental conditions. Field investigations of biofilms usually do not offer the opportunity to study the effects of a large number of experimental variables. The chosen laboratory device should provide the control or determination of a number of chemical, physical and biological parameters that influence the structure, composition and various properties of biofilms (Gjaltema and Griebe, 1995). The following list summarizes experimental variables that are important in the selection, design and construction of biofilm devices when the research question has been defined:

(1) Physical parameters
  - flow regimen
  - flow velocity
  - shear stress
  - temperature
  - surface properties, composition and characteristics
  - hydraulic residence time
(2) Chemical parameters
  - substrate composition and concentration
  - bioproducts in the biofilm matrix and bulk liquid phase
  - redox potential

- inorganic ions
- organic and inorganic particles

(3) Biological parameters
- microorganism type (algae, protozoa, bacteria, viruses, etc.)
- defined or undefined culture
- mixed or pure culture

There are a number of biofilm devices that offer the experimental control of factors influencing the biofilm accumulation. The continuous flow system offers the cultivation for long periods of time and may provide valuable information on the properties of biofilms and the mass balance of substrates/products. The main design philosophy behind the continuous flow system for cultivation of sessile bacteria derived from successful use of the chemostat (Monod, 1950). Only a few devices are commercially available and suitable for a variety of research questions and aims. The RotoTorque™ (rotating annular reactor), constant-depth film fermenters (Atkinson and Fowler, 1974; Peters and Wimpenny, 1988, 1989), radial flow and rotating disc reactor (Fowler and McKay, 1980; Loeb et al., 1984) and the Robbins device (McCoy et al., 1981) are useful devices for the study of biofilm accumulation under defined or controlled hydrodynamic conditions. The Robbins device consists of a tubular container with removable test surfaces (sampling studs). Many experimental systems use the basic principles of the Robbins device for the monitoring of biofilm development, removal and disinfection under plug flow conditions (Banks and Bryers, 1991; Nickel et al., 1985). An overview of the above-mentioned test sytems are described in detail by Gilbert and Allison, (1993).

The RotoTorque™ reactor (Characklis, 1990) was developed from the rotating annular reactor designed by Kornegay and Andrews (1967). The system consists of an inner cylinder that rotates at a constant rate and a stationary outer cylinder. The inner wall of the outer cylinders contains 12 slides that are removable during operation without interruption of flow conditions. The RotoTorque™ has been frequently used for monitoring of biofilm development and resistance toward biocides (Chen et al., 1993a, 1993b; Griebe et al., 1994; Srinivasan et al., 1995). Gjaltema et al. (1994) showed that the RotoTorque™ is of limited use for quantitative physiological or kinetic biofilm studies because the biomass per unit area of sampling slide is not representative of the entire reactor surface area. The Model 920LS (Laboratory) rotating annular reactor system (BioSurface Technologies, Corp., Bozeman, USA) represents a reactor system based on the design of the RotoTorque™. In constrast to the RotoTorque™, the sampling slides of the Model 920LS form a part of the inner cylinder. The Model 920LS rotating annular reactor was used to investigate the distribution of sessile bacteria on various parts of the reactor inner surface. The suitability as a biofilm monitoring system in comparison with the RotoTorque™ is discussed.

MATERIAL AND METHODS

The Model 920LS (Laboratory) rotating annular reactor system (BioSurface Technologies, Corp.) was used for the biofilm experiments (Figure 1). The reactor is essentially a Couette vessel that consists of two concentric cylinders, a stationary outer cylinder (glass) and a rotating inner cylinder (polycarbonate). Twelve removable polished stainless steel slides (SS 304, each 1.9 $\times$ 14.5 cm) form an integral part of the outer wall of inner cylinder (rotating) and permit the sampling of biofilms growing on an area of 27.55 cm^2. This is the most important difference to the original construction design of the RotoTorque™ (Figure 2) where the slides form an integral part of the inside wall of the outer cylinder (stationary) (Figure 3). The bulk liquid is mixed by cylinder rotation and draft tubes (Figure 2) bored through the solid inner cylinder. The rotation of the inner cylinder causes the fluid to rise in the draft tubes, resulting in internal recirculation. The outer and inner cylinders experience a uniform shear stress because the annular gap is constant over the height of the reactor. Because shear stress is uniform and the reactor is well mixed, it is assumed that the biofilm from any of the 12 removable slides (SS304) can be considered as a representative sample. The shear stress is dependent on the rotational speed of the inner cylinder but is independent of the fluid flow rate.

Figure 1 Rotating annular reactor system: Model 920LS (Laboratory) rotating annular reactor system (BioSurface Technologies, Corp.). Disassembled reactor parts: top plate with inlet and bottom plate (left); inner cylinder with stainless steel slide and sample port, outer glass cylinder (middle); assembled reactor system (right).

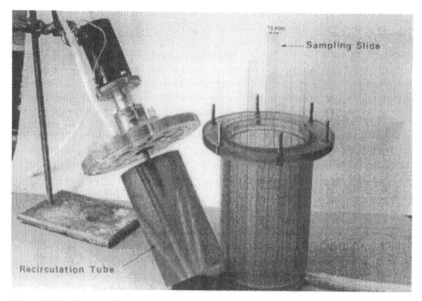

Figure 2 Rotating annular reactor system: RotoTorque™ system (modification after Griebe and Flemming, 1996). Disassembled rotating inner cylinder with draft tubes for recirculation of the bulk liquid and outer cylinder (stationary) with slide.

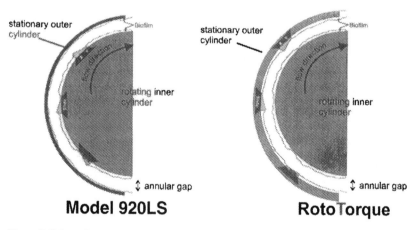

Figure 3 Schematic representation of horizontal section from the two rotating annular reactors: (a) Model 920LS with biofilms on three sampling slides and their position in grooves of the inner rotating cylinder; (b) RotoTorque™ model with biofilms on three sampling slides and their position in grooves of the stationary outer cylinder.

26

Because of the complete mixing of the liquid phase, the rotating annular reactor is equivalent to a continuous flow stirred tank reactor (CFSTR), so the effluent represents the reactor bulk liquid composition.

STERILIZATION AND OPERATION OF THE ROTATING ANNULAR REACTOR

Prior to the experiments the reactor system and silicon tubing were disinfected by biocide treatment. Hydrogen peroxide at 200 mg/l was dosed continuously (30 ml/min) into the reactor for a period of 2 hours. The inner cylinder was rotating at approximately 400 rpm throughout the exposition period to H_2O_2. The reactor effluent was sampled after the disinfection and analyzed by heterotrophic plate count (HPC) on R2A agar (Difco). The efficacy of the disinfection procedure was confirmed by HPC counts of the effluent after the treatment. Thus, it was not necessary to sterilize the reactor system via autoclave at 121°C (15 psi) for 25 min, which may cause deterioration by embrittlement of the material. The sterile rotating annular reactor was fed with drinking water (60 ml/min) corresponding to a residence time of 0.3 hr. The drinking water was supplemented with nutrient broth (40 mg/l, Difco) to promote the growth of an undefined mixed-culture biofilm. The nutrient broth and the drinking water were mixed well prior to the entrance of the reactor to prevent substrate gradients. The inner cylinder was operated constantly at a rotational speed of 300 rpm.

BIOFILM SAMPLING AND ANALYTICAL PROCEDURE

Two metal test surfaces (stainless steel slides, SS 304-) were removed daily and analyzed for bacterial density by using epifluorescence microscopy. The horizontal distribution of sessile bacteria in the rotating annular reactor was analyzed according to the position of the surface (from top to bottom). The upper part of the slide was divided into two sampling areas (A and B). Accordingly, the biofilm of the lower part of the slide was removed from the two sampling areas C and D (Figure 4). At the beginning and at the end of the experiment, one additional slide was analyzed in the vertical direction (X and Y) to investigate the impact of the flow direction on the biofilm density. After 5 days of operation, the reactor was disassembled (see Figure 1), and the biofilm from the areas of the bottom and top plate, and the upper and lower part of the outer cylinder (stationary) were removed and analyzed by microscopic enumeration.

For this purpose the biofilm was stained successively with 4',6-diamidino-2-phenylindole (DAPI, 10 µg/ml, Sigma Chemicals) for 10 min and acridine orange (AO, 0.1 µg/ml, Sigma Chemicals) for 3 min. To avoid the removal of delicate external biofilm structures, the slides were *not* rinsed with sterile

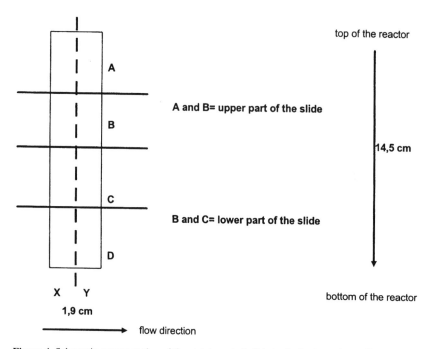

Figure 4 Schematic representation of the stainless steel slide in the horizontal sampling areas A and B of the upper part and C and D of the lower part. X and Y: vertical sampling areas.

filtered water. The originally intended purpose of rinsing is to remove unattached bacteria. However, even standardization of rinsing and washing procedures does not prevent unwanted detachment processes because of the strong forces and content of the rinsing fluids. Dipping of substrata or changing of fluids causes strong shear forces at the air-water interface (Busscher and van der Mei, 1995; Pitt et al., 1993). To prevent unwanted detachment of bacteria, the slide was covered carefully with DAPI and AO solutions. After the staining procedure the slides were fastened on a glass slide and viewed by direct epifluorescence microscopy.

On each slide, approximately 1000 bacterial cells were counted at 1.000× magnification and a micrometer grid of 100 × 100 μm size. After 48 hr of reactor operation the biofilm cell density was too high for direct enumeration on the slide. Indirect enumeration required methods for the removal and disaggregation of biofilm. Sonication and the use of surfactants such as Tween 80 or Triton X-100 proved to be inefficient for the disaggregation of biofilm microorganisms and subsequent microscopic counting of single bacteria after filtration on a membrane filter. The best methods for removal include mechanical scraping with a Teflon scraper and disaggregation with a Teflon piston in a cooled glass cylinder for 5–7 min at 200 rpm (Figure 5). This sam-

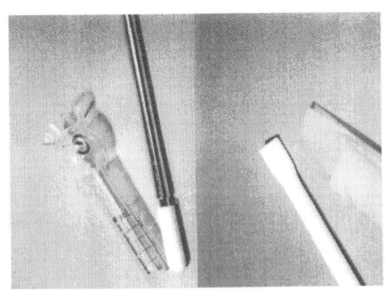

Figure 5 Teflon scraper and piston in a glass cylinder used for removal and disaggregation of biofilms from stainless steel surfaces.

ple preparation gave maximum yield of bacterial cell number and caused no cell lysis. This was confirmed by measurement of the activity of the strictly intracellular enzyme glucose-6-phosphate dehydrogenase (Lessie and Vander Wyk, 1972; Frølund et al., 1995). Cell lysis has to be prevented during biofilm disaggregation because the release of intracellular material causes an increase of disturbing background fluorescence during examination by epifluorescence microscopy.

Total cell counts were determined by counting of stained bacteria with use of an epifluorescence microscope. Disaggregated samples filtered on black polycarbonate membrane filters (diameter 25 mm, pore size 0.2 μm; Milipore) were stained successively with DAPI and AO as described above. At least 20 microscopic grid fields (grid size 100 × 100 μm) or 1000 stained cells were enumerated at 1000× magnification on the polycarbonate membranes. For the statistical analysis the software STATeasy (Lozán, Hamburg-Germany) was used.

RESULTS

The microscopic enumeration of sessile bacteria (undefined mixed-culture) grown on various surfaces and positions in the Model 920LS (Laboratory) rotating annular reactor system (BioSurface Technologies, Corp.) was used for

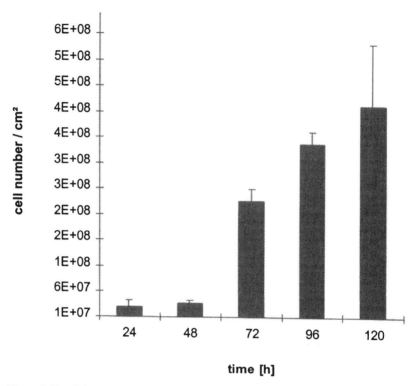

Figure 6 Growth kinetics of biofilm microorganisms on stainless steel slides in the rotating annular reactor (Model 920LS) over a period of 120 hr.

the evaluation of the test system and the suitability as a monitoring device for biofilm accumulation.

Within 1 week on the stainless steel slides, a biofilm was grown with a bacterial density of $4.3 \pm 1.2 \times 10^8/cm^2$. The cell density integrated from all areas (position A, B, C and D) of the slides is shown in Figure 6.

After 48 hr of continuous reactor operation, a dense monolayer of sessile bacteria was observed on the stainless steel slides. A statistical analysis of the enumerated sections (A, B, C and D position) indicated no significant differences between the arithmetic average of normal distribution of numbers of sessile bacteria on the four sections and different sampling slides. A paired Student t-test with a level of significance set at $p < 0.05$ was used for all comparisons. The growth of the sessile bacteria appeared to be representative and homogeneous in horizontal (X and Y position) and vertical direction (A, B, C and D position) on the sampling slides up to a density of 3.5×10^7 cells/cm^2. The direct enumeration of sessile bacteria was in a detection range of 10^4–10^7 cells/cm^{-2} as shown in the schematic diagram (Figure 7).

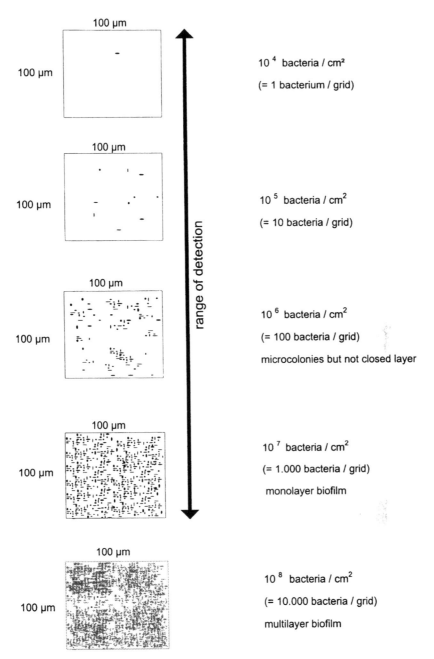

Figure 7 Schematic representation of the detection range for bacterial densities on a surface as viewed through a microscopic counting grid (size 100 μm × 100 μm at 1000× magnification).

TABLE 1.

Sampling Time (h)	Vertical Position Stainless Steel Slide (Upper Part)		Vertical Position Stainless Steel Slide (Lower Part)		Horizontal Position Stainless Steel Slide	
	Position A Cell Number (Mean ± SD/cm^2)	Position B Cell Number (Mean ± SD/cm^2)	Position C Cell Number (Mean ± SD/cm^2)	Position D Cell Number (Mean ± SD/cm^2)	Position X Cell Number (Mean ± SD/cm^2)	Position Y Cell Number (Mean ± SD/cm^2)
24	3.1 ± 1.4 × 10^7	3.5 ± 1.1 × 10^7	3.1 ± 1.0 × 10^7	2.8 ± 1.1 × 10^7	2.9 ± 1.2 × 10^7	2.8 ± 1.5 × 10^7
24	2.8 ± 1.5 × 10^7	2.9 ± 1.4 × 10^7	3.1 ± 1.5 × 10^7	3.1 ± 1.1 × 10^7	3.0 ± 1.2 × 10^7	3.1 ± 1.5 × 10^7
72	1.9 ± 0.17 × 10^8	2.4 ± 0.13 × 10^8	2.9 ± 0.19 × 10^8	3.2 ± 0.41 × 10^8	n.d.	n.d.
72	1.3 ± 0.09 × 10^8	2.1 ± 0.35 × 10^8	2.3 ± 0.29 × 10^8	2.9 ± 0.34 × 10^8	n.d.	n.d.
	Position A + B Cell Number (Mean ± SD/cm^2)		Position C + D Cell Number (Mean ± SD/cm^2)			
96	3.4 ± 0.26 × 10^8		3.3 ± 0.15 × 10^8		n.d.	n.d.
96	2.2 ± 0.21 × 10^8		5.1 ± 0.30 × 10^8		n.d.	n.d.
120	n.d.				5.3 ± 1.2 × 10^8	2.3 ± 0.72 × 10^8
120	n.d.				6.2 ± 1.9 × 10^8	3.1 ± 0.92 × 10^8

Slide

A B C D

Y

X

flow direction

Note: n.d. not determined

32

After 3 days the formation of microcolonies was observed. A multilayer biofilm with stacks was observed by epifluorescence microscopy. Because direct enumeration via epifluorescence microscopy was no longer suitable, the bacterial density was determined after removal from the sampling slide.

The biofilms formed after 3 days exhibited patterns of stripes that appeared to follow the flow direction of the bulk liquid between the rotating inner cylinder and the stationary outer cylinder. The differences in the colonization with bacteria between the upper and lower slide became more pronounced after 4 days of biofilm accumulation (Table 1).

Gradients of bacterial cell numbers in the horizontal (X and Y position) direction of the flow in the annular gap were obvious. After 5 days of reactor operation the biofilm on the slide in flow direction (X position) became consistently thicker and contained significantly higher numbers of bacteria than on the other part of the slide (Y position) (Figure 8). The bacterial density in the X position was with 5.8×10^8 cells/cm^2, two times higher than in the Y position with 2.7×10^8 cells/cm^2.

At the end of the experiment the rotating annular reactor was disassembled and the biofilm was removed from different surfaces for bacterial enumeration (see Figure 1). The distribution of bacteria on the outer cylinder (stationary), the sampling slide from the inner cylinder (rotating), the top plate with the inlet and the bottom plate with the effluent of the rotating annular reactor are shown in Figure 9. The bacterial density of the biofilm on the slide was more than four times higher than on other surface parts of the reactor (Table 2).

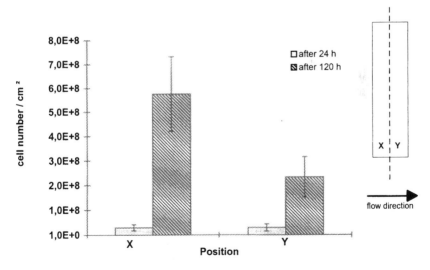

Figure 8 Distribution of bacteria on stainless steel slides in the direction of the flow. Samples were analyzed after 24 and 120 hr of operation.

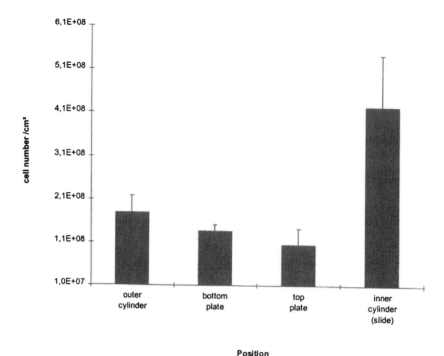

Figure 9 Distribution of bacteria on different surfaces (outer cylinder, glass), (bottom plate and top plate, polycarbonate) and inner cylinder (stainless steel slide) in the rotating annular reactor. Samples were taken after 120 hr of operation.

Similar bacterial cell numbers were found on the outer cylinder (glass), top plate and bottom plate (polycarbonate) despite different construction materials. Consistent with the observation on the sampling slides, a vertical gradient was detected from the upper part (cylinder, top plate) to the lower part (cylinder, bottom plate) of the reactor (Figure 9). A paired Student t-test with a level of significance set at $p < 0.05$ indicated no significant differences between the average bacterial density of the outer cylinder (glass) and the top and bottom plate (polycarbonate); but the density on surface of the upper and lower outer cylinder showed significant differences with $9.8 \pm 2.7 \times 10^7$ cells/cm^2 and $2.6 \pm 0.5 \times 10^8$ cells/cm^2 respectively.

DISCUSSION

Rotating annular reactor systems have been used frequently for quantitative and qualitative biofilm studies (Kornegay and Andrews, 1967; Siebel and Characklis, 1991; Dury et al., 1993; Srinivasan et al., 1995; Kuballa and

TABLE 2.

Position	Outer Cylinder (Upper Part)	Outer Cylinder (Lower Part)	Bottom Plate	Top Plate	Inner Cylinder (Whole Slide)
Material	Glass	Glass	Polycarbonate	Polycarbonate	Stainless Steel
Cell Number (Mean $\pm$ SD/cm^2)	$9.8 \times 10^7 \pm 2.7 \times 10^7$	$2.6 \times 10^8 \pm 5.2 \times 10^7$	$1.4 \times 10^8 \pm 1.4 \times 10^7$	$1.1 \times 10^8 \pm 3.7 \times 10^7$	$4.2 \times 10^8 \pm 1.2 \times 10^8$

35

Griebe, 1995; Griebe and Flemming, 1996; Griebe et al., 1994) because of a variety of physical, chemical and biological parameters that influence the composition, structure, function and properties of biofilms can be controlled or determined under laboratory and under field conditions. It was assumed that a homogeneous and uniform biofilm of constant thickness and biomass developed on all surfaces in the rotating annular reactor. Gjaltema et al. (1994) revealed in an extensive study that the RotoTorque™ was not suitable for the growth of an evenly distributed biofilm with uniform and representative structure and thickness on all reactor surfaces including the sampling surface. They described how the shortcomings of the RotoTorque™ reactor may be obviated by modification of the reactor design and geometry, e.g., the annular gap and the position of the sampling surface and its curvature. The Model 920LS (Laboratory) rotating annular reactor system from BioSurface Technologies, Corp. is essentially a modification of the RotoTorque™. The sampling slides in the Model 920LS do form an integral part of the outer wall of the inner cylinder (rotating), which is the most important difference to the original construction design of the RotoTorque™ (Figure 3) where the slides form an integral part of the inside wall of the outer cylinder (stationary). It was not known whether the changes of the construction design lead to a more evenly distributed biofilm on all reactor surfaces due to the complex flow regimen. Therefore, the reactor system was evaluated by determination of the bacterial densities on different surfaces and positions in the reactor.

Removable flat sampling slides of various materials permit intermittent monitoring and sampling for measuring of biomass and cell number in the RotoTorque™ during operation. This is not possible with the Model 920LS where the slides with the biofilm are rotating during operation. Even slow decrease of the rotational speed may not prevent uncontrolled sloughing because of increasing shear stress on the biofilm. Sloughing events were observed after the growth of a multilayer biofilm, which may explain irregular biofilm pattern on the slide as observed by naked eye and confirmed by high variability of bacterial cell densities between different slides taken from the system at the same time.

The sessile bacteria were homogeneously distributed on the slides up to a cell density of 3×10^7 cells/cm^2. On the sampling area, no gradients in the vertical or horizontal direction of monolayer-forming biofilm microorganisms were detected. The sampling area was representative for the removed slides from the rotating annular reactor. After 3 days of operation, a multilayer biofilm was formed, and differences in the bacterial density on the position of the slides were observed. Horizontal and vertical gradients occurred although the sampling areas (slides) were exposed to a homogeneous shear field. In vertical direction (from the top to the bottom of the reactor), the number of bacteria increased slightly. Visibly thicker biofilms and higher cell densities were detected in the direction of the flow. The results showed a position-dependent colonization of different reactor parts with multilayer biofilms. For qualitative and quantitative studies of defined and representative multilayer biofilms with uniform structure, this type of rotating annular reactor is less suitable, and the results are in close agreement with the findings from Gjaltema et al. (1994).

Despite these shortcomings, the rotating annular reactor is considered as one of the very few available test systems with the ability to control a number of physical, chemical and biological parameters under laboratory and field conditions.

The number of samples available is finite and provides a limited surface area for biofilm removal, which reduces the practicability for quantitative analyses of biofilm formation processes. In the case of the rotating annular reactor Model 920LS, there are 12 removable samples (slides) covering a surface area of more than 330 cm^2. This surface area of the flat slides facilitates an examination of biofilms, which includes light and electron microscopy for counting of bacteria using either direct or indirect methods and morphological observation. Indirect methods for the examination of biomass and composition are applicable because enough biofilm material can be harvested for analysis of, e.g., DNA, carbohydrate, uronic acid, protein, humic substance, TOC and lipopolysaccharide content.

Hydrodynamic characteristics such as flow velocity, flow regimen (laminar or turbulent) and shear stress are important factors influencing mass transfer, detachment rate and structure of the biofilm (Characklis and Marshall, 1990). The hydrodynamics of the rotating annular reactor Model 920LS are controlled by the constant rotational speed of the inner cylinder. Flow velocities and shear stress in the rotating annular reactor have been calculated disregarding the complex flow pattern over the flat slides and the enhanced flow velocity due to the recirculation tubes in the rotating cylinder.

The mixing is closely linked with the type of biofilm device and operational mode: batch or continuous, plug flow or stirred vessel. The rotating annular reactor Model 920LS is considered to be a stirred vessel that can be ideally mixed, but more important than a classification whether a system is ideally mixed or not is to determine the characteristic time for mixing, mass transfer and substrate uptake. In this study the mixing time has not been examined, but the position-dependent bacterial colonisation of different reactor surface indicated a faster substrate uptake than mixing in the reactor, which is consistent with the mixing behavior in the RotoTorque (Gjaltema et al., 1994). The rotating annular reactor Model 920GL offers the simultaneous addition of different types of nutrients and their concentration, pH control, buffer and biocide addition but not direct aeration. This limits the growth of strictly aerobic biofilm microorganisms. The structure and distribution of biofilm in a reactor can be a result of the combination of reactor hydrodynamics, mixing and nutrient supply. In a rotating annular reactor with internal recycle, the total

TABLE 3. Characteristics of the Two Rotating Annular Reactors (RotoTorque™ versus Model 930LS), and, for Comparison, Other Commonly Used Biofilm Devices.

	Rotating Annular Reactors		Rotating Tube	Membrane Reactor	Airlift Reactor	Robbins' Device
	Roto Torque	Model 920LG (Laboratory)				
Flow	Up to 3 m/s, turbulent	Up to 7,5 m/s, turbulent	Low, ill-defined	Size dependent	3-Phase flow, turbulent	Size dependent
Mixing	Ideal, slow	Ideal, slow	Not really	Plug flow (recirculation)	Good	Plug flow: limited
Nutrient Supply	1 Point	1 Point	1 Point	Separated	≥ 1 Point	1 Point
Gradients	Vertical	Horizontal	Yes	Yes, designed	No, biofilm mixed	Yes
Sampling	Convenient, not representative	Not convenient during operation, not representative	Destructive, representative	No/destructive, not representative	Convenient, representative	Not convenient during operation, representative
Sampling Area	360 cm²	336 cm² (1 slide)	Variable	Variable	> 0.5 m², Variable	Size dependent, usually less than 4 cm² (1 stud)
Monitoring	Optical, torque	Optical, torque	Weight	Projecting	No	Optical, pressure
Septic Operation	Yes	Yes	No	Yes	Yes	No
Substratum	Variable, slides	Variable, slides	Variable	Membrane	Variable, size and density limits	Variable, studs
Remarks			Whole device rotates		Collisions	

flow can be ideally mixed, but gradients in biofilm thickness will still occur if the mixing is too slow compared with the substrate conversion or due to local flow areas with insufficient mixing.

The rotating annular reactor Model 920LS is useful for controlled and reproducible generation of thin biofilms under controlled fluid shear stress, independent of bulk residence time with low maintenance and easy handling in laboratory and field applications. The system does not prevent the formation of gradients, which leads to non-representative biofilm samples from the reactor and is therefore not suitable for kinetic studies of multilayer biofilms. However, the rotating annular reactor is considered as one of the very few available test systems with the ability to control a number of physical, chemical and biological parameters under laboratory and field conditions. A list with commonly used biofilm devices and their characteristics are summarized in Table 3.

REFERENCES

Atkinson, B. and H.W. Fowler. 1974. The significance of microbial film fermenters. *Adv. Biochem. Eng.* 3, 224–277.

Banks, M.K. and J.D. Bryers. 1991. Bacterial species dominance within a binary culture biofilm. *Appl. Environ. Microbiol.* 57, 1974–1979.

Busscher, H.J. and H.C. van der Mei. 1995. Use of flow chamber devices and image analysis methods to study microbial adhesion. *Methods Enzymol.* 253, 455–477.

Characklis, W.G. 1990. Microbial fouling control. In: W.G. Characklis and K.C. Marshall (eds.): *Biofilms.* John Wiley, New York, 585–633.

Characklis, W.G. and K.C. Marshall. 1990. *Biofilms.* John Wiley, New York.

Chen, C.-I., T. Griebe and W.G. Characklis. 1993. Biocide action of monochloramine on biofilm systems of *Pseudomonas aeruginosa. Biofouling* 7, 1–23.

Chen, C.-I., T. Griebe, R. Srinivasan and P.S. Stewart. 1993b. Effects of various metal substrata on accumulation of *Pseudomonas aeruginosa.* Biofilms and the efficacy of monochloramine as a biocide. *Biofouling* 7, 241–251.

Dury, W.J., P.S. Stewart and W.G. Characklis. 1993. Transport of 1-μm latex particles in *Pseudomonas aeruginosa* biofilms. *Biotechnol. Bioeng.* 42, 111–117.

Flemming, H.C., T. Griebe and G. Schaule. 1996. Antifouling strategies in technical systems—a short review. *Water Sci. Technol.* 34, 5, 517–524.

Fowler, H.W. and A.J. McKay. 1980. The measurement of microbial adhesion. In: R.W. Berkeley, J.M. Lynch, J. Melling, P.R. Rutter and B. Vincent (eds.): *Microbial adhesion to surfaces.* Ellis Horwood, Chichester, 143–156.

Frølund, B., T. Griebe and P.H. Nielsen. 1995. Enzymatic activity in the activated sludge floc matrix. *Appl. Biotechnol.* 43, 755–761.

Gilbert, P. and D.G. Allison. 1993. Laboratory methods for biofilm production. In: S.P. Denyer, S.P. Gorman and M. Sussman (eds.): *Microbial biofilm: formation and control.* Blackwell Scientific Publications, Oxford.

Gjaltema, A. and T. Griebe. 1995. Laboratory biofilm reactors and on-line monitoring: report of the discussion session. *Water Sci. Technol.* 8, 257–261.

Gjaltema, A., P.A.M. Arts, M.C.M. Loosdrecht, J.G. Kuenen and J.J. Heijnen. 1994. Heterogeneity of biofilms in rotating annular reactors: occurrence, structure, and consequences. *Biotechnol. Bioeng.* 44, 194–204.

Griebe, T. and H.C. Flemming. 1996. Vermeidung von Bioziden in Wasserauf-bereitungs—Systemen durch Nährstoffaufnahme. *Vom Wasser* 86, 217–230.

Griebe, T., C.-I. Chen, R. Srinivasan and P.S. Stewart. 1994. Analysis of biofilm disinfection by monochloramine and free chlorine. In: G.G. Geesey, Z. Lewandowski and H.C. Flemming (eds.): *Biofouling and biocorrosion in industrial water systems.* Lewis Publishers, Boca Raton, FL 151–161.

Kornegay, B.H. and J.F. Andrews. 1967. Kinetics of fixed film biological reactors. *J. Water Pollut. Control. Fed.* 40, R460–R468.

Kuballa, J. and T. Griebe. 1995. Sorption kinetics of tributyl tin on Elbe river biofilms. *Fresenius J. Anal. Chem.* 353, 105–106.

Lessie, T.G. and J.C. Vander Wyk. 1972. Multiple forms of *Pseudomonas multivorans* glucose-6-phosphate and 6-phosphogluconate dehydrogenases: differences in size, pyridine nucleotide specificity, and susceptibility to inhibition by adenosine 5′-triphosphate. *J. Bact.* 110, 3, 1107–1117.

Loeb, G.I., D. Laster and T. Gracik. 1994. The influence of microbial fouling films on hydrodynamic drag of rotating discs. In: J.D. Costlow and R.C. Tipper (eds.): *Marine biodeterioration: an interdisciplinary study.* Naval Research Press, Annapolis, MD, 88–100.

McCoy, W., J.D. Bryers, J. Robbins and J.W. Costerton. 1981. Observations in fouling biofilm formation. *Can J. Microbiol.* 27, 910–917.

Monod, J. 1950. La technique de culture continué.Théorie et applications. *Annales de l'Institut Pasteur* 79, 390–409.

Nickel, W.M., I. Ruseska, J.B. Wright and J.W. Costerton. 1985. Tobramycin resistance of cells of *Pseudomonas aeruginosa* growing as a biofilm on urinary catheter material. *Antimicrob Agents Chemother* 27, 619–624.

Peters, A.C. and J.W.T. Wimpenny. 1989. A constant depth laboratory model film fermenter. In: J.W.T. Wimpenny (ed.): *Handbook of laboratory model systems for microbial ecosystems research.* CRC Press, Boca Raton, 175–195.

Peter, A.C. and J.W.T. Wimpenny. 1988. A constant depth laboratory model film fermenter. *Biotechnol. Bioeng.* 32, 263–270.

Pitt, W.G., M.O. McBride, A.J. Barton and R.D. Sagers. 1993. Air-water interfaces displace adsorbed bacteria. *Biomaterials* 14, 605–608.

Siebel, M.A. and W.G. Characklis. 1991. Observations of binary population biofilm. *Water Res.* 19, 1369–1378.

Srinivasan, R., P.S. Stewart, T. Griebe, C.-I. Chen and X. Xu. 1995. Biofilm parameters influencing biocide efficacy. *Biotechnol. Bioeng.* 46, 553–560.

CHAPTER 3

Microcalorimetry for Evaluating Countermeasures against Biofouling in Water Circulation Systems

HENRY VON RÈGE
WOLFGANG SAND

INTRODUCTION

WATER circulation systems like heat exchangers or cooling systems are often affected by biofouling and MIC (microbially influenced corrosion) (Donlan et al., 1994; Flemming, 1995). Biofilms are ubiquitous in water systems and form on every surface. The undesired effects are corrosion of construction materials, heat loss in heat exchangers, or the decrease of product quality (paper industry). Therefore, the operators have to keep their systems free from biofouling. Because the costs for countermeasures are high, there is obviously a need for effective methods to evaluate these. One of the most common ways to eliminate biofouling is the use of biocides. However, the manufacturer's recommendation for dosage often needs to be modified, because biofilm microorganisms are more resistant to biocide action than planktonik ones. As a result, the dose recommendation is often too low to inhibit microbial activity and/or to kill the microorganisms in the biofilm.

To solve these problems, microcalorimetry may be used as a treatment to control against biofouling and MIC. Starting in the 1970s this technique has been successfully used in medical research to optimize the choice and dosage of antibiotics (Binford et al., 1973). Furthermore, microcalorimetry has been used for monitoring microbial growth (von Stockar and Marison, 1989), cell adhesion to surfaces (Humphrey and Marshall, 1984), and biofilm generation (Wentzien et al., 1994a). Different habitats like soil (Ljungholm et al., 1979), solid surfaces such as ore (Schröter and Sand, 1993) or biofilms (Lock and Ford, 1983) can easily be analyzed for microbial activity. The metabolism of either aerobic or anaerobic bacteria can be measured (Traore et al., 1981; von Stockar and Marison, 1989). Microcalorimetry is also well suited for measurements of slow growing, chemolithotrophic bacteria (Schröter and Sand, 1993). The substrate turnover can be determined with calorimetric measurements as shown for bacterial pyrite oxidation (Rohwerder et al., 1998). Thus, monitoring of acid rock drainage formation in sulfidic mine waste and of leaching activity in bioleaching plants for metal recovery becomes possible (Rohwerder et al., 1997; Schippers et al., 1995).

The microbial activity in complex systems is measured by the detection of heat loss that accompanies all biochemical redox reactions. Heat quantities as small as 10^{-6} W, e.g., evolved by bacteria, may be recorded. Heat, which is released from biochemical reactions, flows from the sample to a large heat sink. Peltier elements surround the heat-producing sample and convert the heat flow into a voltage signal. In case of surface-attached bacteria the limit of detectability is in the range of $10^4-10^5/cm^2$ calculated from results of Bettelheim and coworkers (Bettelheim and Shaw, 1980). For leaching bacteria attached to ore it is documented that just 10^2 cells/g ore are able to produce a measurable amount of heat (Wentzien et al., 1994b).

In recent years, a microcalorimetric test was developed to determine microbial activity of biofilm samples on metals. A major advantage of this test is that the biofilm remains on the surface or does not need a further treatment for the measurement. Thus, this non-destructive technique enables real on-line measurements of microbial activity in the biofilm. The capability for testing biocide treatment and for screening and optimizing countermeasures against biofouling and MIC has been demonstrated previously (von Rège and Sand, 1996a; 1998).

In this chapter we present how this test may be used to determine biological and chemical activity in biofilm samples on mild and stainless steel and how to evaluate which method effectively reduces microbial activity and viable cells in biofilms.

MATERIALS AND METHODS

CALORIMETER

A thermal activity monitor (TAM 2277, Thermometric AB, Broma, Sweden; or C3-Analysentechnik, Baldham, Germany) was used for the isothermal heat conduction measurements (Wadsö, 1974). The calorimeter was equipped with an ampule-cylinder (Nr. 2277-205) for measuring heat output of biofilm samples on metal coupons. Measurements were performed in a 25-ml stainless steel ampule. For flow-through experiments a flow-through measuring cylinder (Nr. 2277-202) was used. The cylinder contains gold tubing mounted on the measuring cup. Experimental procedure and setup were performed as described elsewhere (Wentzien et al., 1994a).

DETERMINATION OF BIOLOGICAL AND CHEMICAL ACTIVITY OF BIOFILM SAMPLES ON METAL COUPONS

The 25-ml stainless steel ampule was sterilized by immersion in 70% ethanol for 15 min prior to a measurement. To avoid a reoxidation of reduced compounds, in the case of anaerobically incubated biofilm samples, the following steps, after withdrawal of sample coupons, were performed under anaerobic conditions in an anaerobic box (du-scientific, USA). Samples that originated from experiments conducted under the presence of oxygen were processed without anaerobic sampling procedure. Immediately after withdrawal of a biofilm coupon, it was rinsed with deionized water and inserted into an ampule. The ampule was filled with 15 ml of sterile nutrient-medium (von Rège and Sand, 1996b; supplemented with 3.5 g/l lactate and 1.0 g/l yeast extract to guarantee full nutrient supply). The closed ampule was allowed to equilibrate at 30°C for 75 min before starting the measurement. The ampule was lowered in the measuring position and recording was started. The signal for heat output was recorded via a computer (software Digitam 4.01, Thermometric, Sweden). The heat output of a sample was determined after obtaining a stable signal that was established within 2 hr. Afterward, the biofilm sample was incubated with a 5% formaldehyde solution for 24 hr to determine the biocidal effect. The same sample was then measured again to determine the (chemical) heat output. The difference between both values was calculated as microbial (biological) activity. To evaluate the biocide efficacy on biofilm organisms, the coupons were analyzed for cell counts of bacteria by the MPN technique. Cell counts of bacteria at the start of the experiments before incubation with formaldehyde and the weight loss caused by corrosion were performed by using parallel coupons.

FLOW-THROUGH EXPERIMENTS

The flow-through cylinder contains gold tubing mounted on the measuring cup. By pumping a solution with growing bacteria continuously through the tubing, the heat output caused by metabolism can be measured by the surrounding Peltier elements. If bacteria do attach in the tubing and form a biofilm, microbial activity can thus be determined.

A pure culture of *Vibrio natriegens* (DSM 759) was cultivated in an external incubator (0.5 g/l peptone, 0.1 g/l meat extract, 0.2 g/l yeast extract, 1.21 g/l Tris, 25.0 g/l NaCl, pH 7.5, 30°C, aerated). During growth, the culture solution was continuously pumped through the flow-through cylinder. After 5 hr of pumping the culture solution through the tubing, it was substituted by sterile medium, to detect a possible biofilm formation on the surface of the gold tubing. After reaching a stable heat output value, indicating a mature biofilm in the tubing, the cylinder was rinsed for 7 hr with an SDS/NaOH-solution (0.1 M NaOH/0.1% SDS) to disrupt the biofilm. Afterward, sterile medium was pumped through the cylinder, to determine whether there was still microbial activity resulting from surviving cells or perhaps regrowth.

In a second set of experiments the action of a commercially available bio-cide (Dilurit, BEK Ladenburg, Germany) on biofilms was investigated. After establishing a biofilm of *Vibrio natriegens* on the surface of the tubing in the cylinder, the biocide was added at a concentration of 500 ppm. After biocide treatment, the medium was exchanged against sterile biocide-free medium, and activity was measured to monitor a possible regrowth of biofilm bacteria.

MICROORGANISMS AND BIOFILM GROWTH ON METAL COUPONS

Biofilms on metal coupons (surface of 12 cm^2) of mild (St 37) or stainless steel (AISI 304) were produced in a miniplant. The miniplant consisted of ring columns containing internal coupons and a nutrient reservoir, where O$_2$, pH, redox potential, and temperature were controlled (von Rège and Sand, 1996a, b). Experiments were conducted with *Thiomonas intermedia* K12 (formerly *Thiobacillus intermedius*), *Desulfovibrio vulgaris* NCIMB 8457, and *Ochrobactrum anthropi* (von Rège and Sand, 1996b). Bacteria were cultivated in a modified Postgate medium C with 100.0 mg/l lactate, 50.0 mg/l yeast ex-tract, and 5.0 g/l sodium thiosulfate. The latter was added to support the growth of *Thiomonas intermedia*. Fresh medium was added continuously to the reser-voir (flow rate 480 ml/day). The pH in the circulating medium was kept at 7.0 to establish *Thiomonas intermedia* in the biofilm community. Experiments were conducted for 6 weeks at 30°C under either aeration or N$_2$ (anaerobic conditions) or alternating aeration (change between aerobic and anaerobic con-ditions). The biofilm was analyzed for cell counts by MPN and for microbial (biological) and chemical activity by microcalorimetry.

RESULTS AND DISCUSSION

DETERMINATION OF MICROBIAL ACTIVITY OF BIOFILM ON METAL SAMPLES

The aim of these experiments was to estimate the part of biological and chemical heat output produced by a biofilm on metallic materials. Further-more, the efficacy of biocide treatment (5% formaldehyde) was investigated.

The highest heat output from biofilm samples was 30 µW/cm^2, originating from mild steel coupons that had been incubated under alternating aeration. The biofilm consisted of *Thiomonas intermedia, Desulfovibrio vulgaris* and *Ochrobactrum anthropi* with cell counts of about 10^6/cm^2 (Figure 1). After

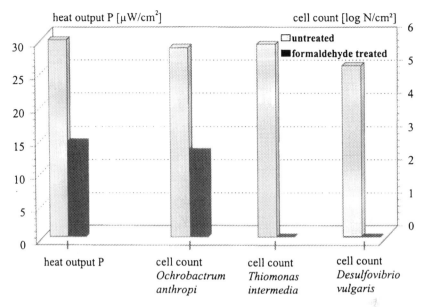

Figure 1 Heat output and cell counts of biofilm samples on mild steel coupons incubated for 6 weeks under alternating aeration with a mixed culture of *Thiomonas intermedia, Desulfovibrio vulgaris* and *Ochrobactrum anthropi* in a miniplant. Samples were measured before and after an incubation for 24 hr in 5% formaldehyde solution.

incubating these biofilm samples with the formaldehyde solution for 24 hr the heat output decreased to 14 μW/cm^2. The biofilm consisted only of *Ochrobactrum anthropi* with 10^2 cells/cm^2. Cells of *Thiomonas intermedia* and *Desulfovibrio vulgaris* were no longer detectable. For microcalorimetric investigations cell numbers of at least 10^4–10^5/ml are considered as a limit of detectability (Bettelheim and Shaw, 1980). Thus, the heat output after formaldehyde addition originated only from chemical activity in the biofilm like the chemical oxidation of reduced sulfur compounds or the oxidation of the metal. Thus, the biological activity of the biofilm samples amounted to 50% of the total measured activity. In contrast, the total heat output of stainless steel samples was only 5 μW/cm^2. After incubation with formaldehyde the heat output decreased to 0.6 μW/cm^2 (Figure 2). Again, only cells of *Ochrobactrum anthropi* were detectable, and the cell number was too low to produce a measurable heat output. In this case, however, the biological activity amounted to 90%. Biofilm samples on mild and stainless steel coupons incubated under aerobic or anaerobic conditions exhibited a similar pattern of results (not shown). On mild steel the chemical activity generally contributed more to the total activity, due to corrosion effects, than on stainless steel coupons (without corrosion).

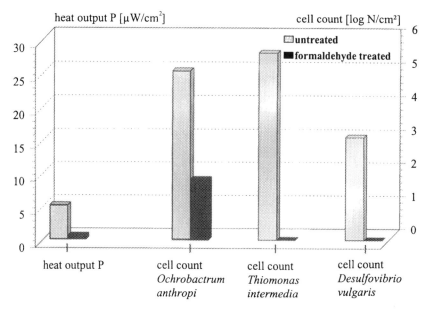

Figure 2 Heat output and cell counts of biofilm samples on stainless steel coupons (ASI 304) incubated for 6 weeks under alternating aeration with a mixed culture of *Thiomonas intermedia, Desulfovibrio vulgaris* and *Ochrobactrum anthropi* in a miniplant. Samples were measured before and after an incubation for 24 hr in 5% formaldehyde solution.

The effect of chemical agents, biocides, or cleaning procedures may easily be quantified by this activity test. Previous experiments have demonstrated that treatment with a chloroform or a water/methanol/chloroform solution did not quantitatively kill biofilm microorganisms. As a result, some biological activity remained detectable (results not shown). Nevertheless, we were able to show how this technique may be used for screening and optimization of biocide treatment against biofilm bacteria (von Rège and Sand, 1996a; von Rège and Sand, 1998). It was demonstrated that the dosage of a commercial biocide has to be considerably higher to inhibit microbial activity as recommended by the producer and to reduce cell numbers in the biofilm effectively. Further experiments to elaborate a routine method based on this test procedure are in progress.

BIOCIDE AND CLEANING TREATMENT ON MICROBIAL ACTIVITY IN BIOFILM

Experiments with *Vibrio natriegens* were performed to investigate if biofilm generation on the surface of the gold tubing in the flow-through cylinder occurred with this organism and whether cleaning procedures and biocide action can be determined as efficiently as with the previously used coupons in the ampoule.

A growing culture of *Vibrio natriegens* was circulated through the flow-through cylinder for 5 hr. Afterward, the culture solution was substituted with fresh, sterile medium. The heat output after substitution increased to a maximum of 170 μW and finally reached a stable value at 75 μW (Figure 3). The heat output is caused by microbial activity of cells growing in a biofilm in the gold tubing of the flow-through cylinder. The shape of the curve represents, first, the growth of the biofilm to a maximum and, second, a stationary phase (1). At this time the cylinder was cleaned with an NaOH/SDS-solution to disrupt and detach the biofilm (2). Seven hours later, sterile medium was pumped through the cylinder to test whether a regrowth of surviving organisms might occur. Over a period of 10 hr (3) the heat output did not increase, demonstrating the efficacy of the cleaning procedure.

Because the results demonstrated the possibility of using this type of assay for screening the efficacy of a biocidal treatment, another experiment was run with a commercially available biocide. The stationary phase of this experiment exhibited a heat output of 100 μW (Figure 4). At this time (2) the biocide was added at a concentration of 500 ppm. The heat output decreased rapidly to 1–2 μW and remained at this value. After exchange against sterile medium without biocide, the heat output increased within 8 hr and reached the same value as before biocide addition. Thus, with this assay an optimization of biocide dosage and incubation time is possible. It can be more sensitive than the coupon/ampule procedure. It may also be used for biocide screen-

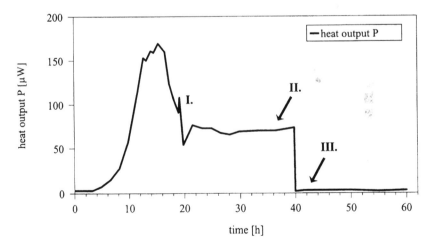

Figure 3 Heat output of a biofilm culture of *Vibrio natriegens* established on the surface of gold tubing in a flow-through cylinder in the microcalorimeter. Phase I: biofilm formation in the tubing of the cylinder, followed by application of sterile medium. Phase II: removal of biofilm by rinsing the tubing for 7 hr with an NaOH/SDS-solution. Phase III: test for detection of remaining living biofilm cells by supplying fresh, sterile medium.

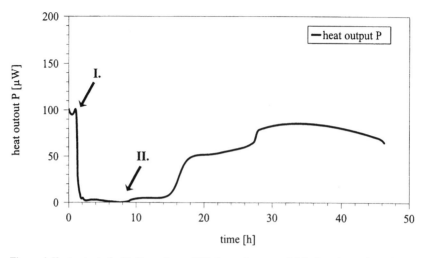

Figure 4 Heat output of a biofilm culture of *Vibrio natriegens* established on the surface of gold tubing in a flow-through cylinder in the microcalorimeter. Phase 1: addition of 500 ppm biocide (Dilurit) after establishment of a stable biofilm. Phase 2: regrowth of remaining living biofilm cells, after exchange of biocide-containing medium against a biocide-free one.

ing, where the impact of biocides against different groups of bacteria needs to be tested. By replacing the tubing material (metals, plastics, etc.) the materials resistance against bacterial attachment can be investigated and real on-line measurements of microbial activity of biofilm can be performed.

CONCLUSIONS

Biological and chemical activity of biofilm can be detected and differentiated on coupons of mild and stainless steel by the short-time activity test. The advantages of this fast and non-destructive technique for an evaluation of countermeasures against biofouling (and MIC) was demonstrated. By the use of this test a biocide action against undesired biofilms in water circulation systems may be optimized, allowing to identify the optimal dosage and incubation time with relation to the relevant bacterial community and the construction material of the system in question. Samples can be obtained by the use of the transportable miniplant to be connected to the tested system. Different materials can simultaneously be investigated. Biofilm samples are analyzed for corrosion/fouling relevant microorganisms, microbial activity and corrosion resistance. Thus, microbial activity in such systems can be monitored and controlled. With the flow-through equipment real on-line measurements of microbial activity in biofilm can be performed, allowing for biocide screening with increased sensitivity. Further experiments for opimizing both systems to screen different materials for an endangerment by biofilm and biofouling are in progress.

ACKNOWLEDGEMENT

This study is based in part on a doctoral thesis of Henry von Rège in the faculty of Biology, University of Hamburg. It was supported by grant AIF 9524 and 10653 N/1 to W.S. through BMWi.

REFERENCES

Bettelheim, K.A. and E.J. Shaw. 1980. *Biological microcalorimetry.* Academic Press, London, 187–194.

Binford, J.S., L.S. Binford and P. Adler. 1973. A semiautomated microcalorimetric method of antibiotic sensitivity testing. *Am. J. Clin. Pathol.* 59, 86–94.

Donlan, R.M., T.L. Pipes and T.L. Yohe. 1994. Biofilm formation on cast iron sub- strata in water distribution systems. *Water Res.* 28, 1497–1503.

Flemming, H.-C. 1995. Biofouling and biocorrosion—effects of undesired biofilms. *Chemie/Ingenieur/Technik* 67, 1425–1430.

Humphrey, B.A. and K.C. Marshall. 1984. The triggering effect of surfaces and sur- factants on heat output, oxygen consumption and size reduction of a starving ma- rine *Vibrio. Arch. Microbiol.* 140, 166–170.

Ljungholm, K., B. Noren, R. Skold and I. Wadsö. 1979. Use of microcalorimetry for the characterization of microbial activity in soil. *OIKOS* 33, 15–23.

Lock, M.A. and T.E. Ford. 1983. Inexpensive flow microcalorimeter for measuring heat production of attached and sedimentary aquatic microorganisms. *Appl. En- viron. Microbiol.* 46, 463–467.

Rohwerder, T., A. Schippers and W. Sand. 1997. Proceedings of the XX International Mineral Processing Congress, Aachen, Vol. 4, Solid-liquid Separation, Hydro- und Biohydrometallurgy, Gesellschaft für Bergbau, Metallurgie, Rohstoff- und Umwelttechnik (GMDB), Clausthal-Zellerfeld, Germany, 475.

Rohwerder, T., A. Schippers and W. Sand. 1998. Determination of reaction energy val- ues for biological pyrite oxidation by calorimetry. *Thermochimica Acta* 309, 79–85.

Schippers, A., R. Hallmann, S. Wentzien and W. Sand. 1995. Microbial diversity in uranium mine waste heaps. *Appl. Environ. Microbiol.* 61, 2930–2935.

Schröter, A.W. and W. Sand. 1993. Estimations on the degradability of ores and bac- terial leaching activity using short-time microcalorimetric tests. *FEMS Microbiol. Rev.* 11, 79–86.

Traore, A.S., C.E. Hatchikian, J. Le Gali and J.-P. Belaich. 1981. Microcalorimetric studies of the growth of sulfate-reducing bacteria: energetics of *Desulfovibrio vul- garis* growth. *J. Bacteriol.* 145, 191–199.

von Rège, H. and W. Sand. 1996a. Simulation of metal-MIC for evaluation of coun- termeasures. *Materials and Corrosion* 47, 486–494.

von Rège, H. and W. Sand. 1996b. *Biodeterioration and biodegradation*. DECHEMA Monographs, Vol. 133. WILEY-VCH, Weinheim, Berlin, 325–331.

von Rège, H. and W. Sand. 1998. Evaluation of biocide efficacy by microcalorimetric determination of microbial activity in biofilms. *J. Microbiol. Methods*. 33, 227–235.

von Stockar, U. and W. Marison. 1989. *Advances in biochemical engineering/biotechnology*. Springer-Verlag, Berlin-Heidelberg, Germany, 94–136.

Wadsö, I. 1974. A microcalorimeter for biological analysis. *LKB J*. 21, 18–21.

Wentzien, S., W. Sand, A. Albertsen and R. Steudel. 1994a. Thiosulfate and tetrathionate degradation as well as biofilm generation by *Thiobacillus intermedius* and *Thiobacillus versutus* studied by microcalorimetry, HPLC, and ion-pair chromatography. *Arch. Microbiol*. 161, 116–125.

Wentzien, S., R. Hallmann and W. Sand. 1994b. *Fachtagung Überwachungsmethoden*, Gentechniktexte 29/94. Umweltbundesamt, Berlin, Germany, 297–304.

CHAPTER 4

Extraction and Quantification of Extracellular Polymeric Substances from Wastewater

ROSEMARIE SPAETH
STEFAN WUERTZ

INTRODUCTION

In past decades different methods for the extraction of extracellular poly-meric substances (EPS) from biofilms and activated sludge have been de-veloped. The cause for this increased research activity has been a growing interest in the composition and function of extracellular biopolymers and their influence on the properties of biofilms and activated sludge.

A definition of EPS is both straightforward and difficult because of historical considerations. There are numerous studies dealing with the isolation and identification of specific biopolymers, usually of a certain commercial value. Largely based on pure culture work they were long thought to contain only polysaccharides. EPS were defined as those polysaccharides that were associated with the cell either in the form of a tightly bound capsule or as loosely associated polymers. It is now generally recognized that EPS contain more than polysaccharides, and we will define them here in the broadest possible sense. In keeping with the suggestions of other workers (Flemming et al., 1996; Nielsen et al., 1997; Flemming, 1995; Costerton and Irvin, 1981) EPS are considered to contain polysaccharides, proteins (including enzymes), DNA, lipids and uronic acids but not lipopolysacccharides, teichoic acid, lipoteichoic acid and teichuronic acid, or crystalline S-layer.

Capsules are composed of EPS closely associated with the cell envelope and behave like cells in negative staining procedures, i.e., they exclude the stain. The opposite is true for more loosely bound polymers (Flemming et al., 1996). Because capsules can impart a certain resistance to shear forces on cells (Pedersen, 1990) and create a buffer zone around them, they may be essential to the survival of cells in nature (Geesey and Jang, 1989; Flemming et al., 1996). In the medical literature capsules protect bacterial cells and play an important

51

role in pathogenicity (Brisou, 1995). In this context we do not support the claim by Gehr and Henry (1983) who defined extracellular polymeric substances as "that material which can be removed from microorganisms (and in particular, bacteria) without disrupting the cell, and without which the microorganism is still viable" (p. 1743). However, certain components of capsules may not be necessary for survival of pure cultures maintained on rich growth media.

A fair amount of confusion has been spread in the literature, both scientific and popular, by the use of the term "slime." Although it is obvious that EPS thus described cannot be part of the capsular EPS, the differences between "slime" and loosely associated polymers are not always clear. In this chapter we will apply the definition recently proposed by some workers (Hsieh et al., 1994; Nielsen et al., 1997; Nielsen and Jahn, 1999). *Bound* EPS is meant to include both capsular and loosely bound EPS and any other attached material; *soluble* EPS applies to "slime," colloidal material and soluble macromolecules (Nielsen and Jahn, 1999). In practice the soluble EPS is usually removed from cell aggregates in a washing step prior to extraction proper and thus does not come under the definition of extracted biopolymers (see below). The term "glycocalyx" has been used to include both bound and soluble EPS, and we will treat it as being synonymous with extracellular polymeric substances. The reader should note that any such definition of EPS must necessarily be an operational one and is open to criticism.

From the point of view of the organisms surrounded by them, EPS are composed of macromolecules produced and secreted by cells and cellular debris and the products of extracellular hydrolytic activity. In addition, EPS can contain a variety of sorbed substances including pollutants and incorporated particulate material. Within a biofilm, EPS form a protective layer that cushions microorganisms against adverse environmental changes such as desiccation, pH value extremes, salt exposure, biocides and hydraulic pressure. Biofilms thus equipped are able to withstand the most extreme conditions (Flemming, 1991). Activated sludge too can be described as a suspension of flocs in which living and dead microorganisms, together with inorganic material, are embedded in a polymeric matrix (Sanin and Veslind, 1994). In either case the presence of EPS influences the specific properties of biomass. In this chapter we will limit ourselves to a discussion of EPS as they occur in mixed culture systems, namely, activated sludge and biofilms. Those readers interested in a more general treatise of extraction methods, including those pertaining to pure cultures, are referred to the recent review by Nielsen and Jahn (1999).

An array of techniques has been applied to the investigation of structure and composition of biofilms and activated sludge flocs. TEM and SEM have illustrated both spatial distribution and structure of cells (Surman et al., 1996). They have the disadvantage of needing sample preparation such as dehydration and fixation resulting in a loss of three-dimensional architecture. In addition, both methods have innate inaccuracies in the form of possible artifacts that can be introduced. For these reasons invasive techniques have proven to be unsatisfactory for the direct observation of EPS.

There are a number of different methods that allow the direct and noninvasive detection of biological matrices as summarized by Nivens et al. (1995). They include fluorescence microscopy and specific fluorophores to obtain information about structure and composition of biofilms and flocs. Calcofluor White is a stilbene dye reported to be relatively specific for cellulose and other β-linked polysaccharides and which has been used to visualize EPS (Del Gallo et al., 1989; Spaeth, 1998); lectins are a vast group of glycoproteins first discovered in 1888 (Stillmark, 1888) that recognize specific patterns or sequences in carbohydrates and attach to them. They are widespread in nature and occur in bacteria, plants, animals and humans (Brisou, 1995). More than a hundred lectins have been commercialized and are available with fluorescent labels. Lectins have been used for the detection of extracellular polysaccharides in conjunction with confocal laser scanning microscopy (CLSM) (Neu, 1996). Because of the specificity of the lectins used, it is possible to describe both the spatial distribution and the composition of some components of EPS. CLSM has also been used to investigate sorption processes in biofilms. Wolfaardt et al. (1995) revealed the accumulation of the insecticide diclofop methyl and its degradation products directly with CLSM and without destroying the biofilm structure. Microscopically based methods allow for a qualitative characterization but have not been used to quantitate EPS.

Special emphasis has been placed on the reaction of biofilm or activated sludge to changing environmental conditions including temperature and pH changes, substrate availability and the presence of toxic substances. These changes can lead to an increase or a reduction in the amount of EPS in biomass. For example, it was shown by Fourier transform infrared spectroscopy (FTIR) that toluene stimulated the production of polysaccharides and, at higher concentrations, the production of carboxylic groups (Schmitt et al., 1995).

One way to detect all biopolymers and, hopefully, quantify them, consists of separating EPS from cells, which is the subject of this chapter. This approach has been used in biological wastewater treatment to investigate sorption properties (Brown and Lester, 1982; Wuertz et al., submitted) and dewatering of activated sludge (Sanin and Vesilind, 1994). Extraction of EPS enables the quantification of EPS content in the biomass and the determination of its constituents. This information can then be used to better comprehend the properties of activated sludge and biofilms.

EXTRACTION METHODS

The architecture and microbial population structure of activated sludge and biofilms are highly complex. EPS play a major role in determining the characteristics of biofilm and activated sludge properties based on their distinct physicochemical properties; the chemical composition of EPS is responsible for cohesiveness and mechanical stability. Between individual EPS components there will be an array of weak interactive forces such as van der Waals forces, electrostatic forces, or hydrogen bonds. Another possibility for the internal stability of cell aggregates is the influence of divalent cations like calcium and magnesium (Bruus et al., 1992), which can form bridges between polysaccharide chains and thus enable the formation of a gel-like matrix. EPS contain both hydrophilic and hydrophobic substances including proteins and lipids (Urbain et al., 1993). Hence, there can be interactions between these hydrophobic substances and the cell surface, thereby reinforcing binding of EPS to cells.

A multitude of methods for the extraction of polymers have been developed on the basis of the destruction of these binding forces. Both quality and quantity of separated EPS extracts depend on the strength of the interactions between extracellular biopolymers and the cell surface. The basis of most methods is formed by procedures developed for the extraction of EPS from pure cultures. However, these methods are not directly applicable to mixed cultures such as activated sludge or biofilms (Brown and Lester, 1980). The overwhelming majority of existing methods is based on washing, extraction and quantification steps. Table 1 summarizes some of these procedures. The washing step is meant to separate soluble biopolymers from the biomass by low-speed centrifugation. These substances are commonly referred to as "slime." In general, this fraction is not considered as part of the EPS in activated sludge flocs and biofilms.

The second step in the determination of EPS is the removal of more strongly bound biopolymers from cell aggregates and the cell surface. The methods in use can be divided into those based on physical principles of separation and those based on chemical procedures. The main steps of physical techniques for the extraction of EPS are high-speed centrifugation, ultrasound treatment and homogenization, or thermal treatment. To achieve a better separation of extracellular polymers, several procedures include the application of certain chemical compounds that allow the destruction of floc architecture. Examples are acid hydrolysis with HCl or H_2SO_4, alkaline conditions or the use of organic solvents. Most methods use centrifugation to physically separate EPS from cell material with speeds varying greatly from $6.000 \times g$ to $32.500 \times g$ (Horan and Eccles 1986; Pavoni et al., 1972). Despite these efforts, it has been shown that the speed of centrifugation does not have a significant effect on the efficiency of EPS extraction (Novak and Haugan, 1981; Platt and Geesey, 1985).

Another approach consists of complexing divalent ions such as calcium and magnesium based on the assumption that the major binding mechanism among EPS molecules is bridging of electrostatic interactions by divalent cations. By

TABLE 1. Methods for the Extraction of Extracellular Polymeric Substances from Different Biological Matrices.

Source of EPS	Method of EPS Extraction	Extraction Products	Reference
Activated sludge	Centrifugation, ethanol precipitation	Proteins, RNA, carbohydrates, DNA	Pavoni et al., 1972
Activated sludge	Alkaline extraction, ethanol precipitation	Carbohydrates, DNA	Nishikawa, and Kuriyama, 1968
Cells of freshwater sediment bacterium	EDTA, ethanol precipitation	Carbohydrates, DNA, RNA	Platt and Geesey, 1985
	EDTA	Proteins, carbohydrates, uronic acids, DNA	
Activated sludge	Ion exchange resin, NaOH, thermal treatment	Proteins, uronic acids, carbohydrates, humic acid	Frølund et al., 1996
Activated sludge biofilm	EDTA, ion exchange resin, crown ether	Proteins, uronic acid, carbohydrates	Spaeth et al., submitted
Activated sludge	Thermal treatment, ion exchange resin, NaOH ultrasonication, homogenization	Proteins, carbohydrates	Rudd et al., 1983
Activated sludge	Sonication	Proteins, DNA, carbohydrates	Urbain et al., 1993
Activated sludge	Thermal treatment	Proteins, carbohydrates, lipids	Goodwin and Forster, 1985
Anaerobic and activated sludge	Thermal treatment, acetone/ethanol precipitation	Proteins, carbohydrates	Morgan et al., 1990
Activated sludge, Klebsiella aerogenes	High-speed centrifugation, ultrasonication, steaming, NaOH, EDTA	Proteins, carbohydrates, DNA, uronic acids	Brown and Lester, 1980
Anaerobic sludge	Ion exchange resin, NaOH thermal treatment, phenol treatment, ethanol precipitation	Proteins, carbohydrates	Karapanagiotis et al., 1989
Activated sludge	High-speed centrifugation, thermal treatment combined with acid or alkaline, blending, potassium-biphosphate	Ethanol-insoluble material	Gehr and Henry, 1983
Activated sludge	Centrifugation, glutaraldehyde extraction	Proteins, carbohydrates	Azeredo et al., 1998
Activated sludge, Escherichia coli	Alkaline extraction, ethanol precipitation	Carbohydrate, RNA, DNA, proteins	Sato and Ose, 1980

removing bridging metal ions the stability of the EPS matrix is weakened, and a physical separation of EPS by shear forces and centrifugation is facilitated. To complex divalent cations investigators have used ethylenediaminetetraacetic acid (EDTA) (Platt and Geesey, 1985) and a crown ether selective for alkaline and alkaline earth metals (Wuertz et al., submitted). Another option for the extraction of EPS is the use of ion exchange resins (Frølund et al., 1996). Glutaraldehyde has been used successfully to solubilize EPS constituents (Azeredo et al., 1998). The yield was comparable with that using the cation exchange resin (Frølund et al., 1995) and little cell lysis was observed. The exact mode of action of glutaraldehyde during extraction is not understood (Azeredo et al., 1998).

Other workers have compared methods for activated sludge (Brown and Lester, 1980; Gehr and Henry, 1983; Rudd et al., 1983) and anaerobically digested sludge (Karapanagiotis et al., 1989). These investigations showed that centrifugation alone removes only a very small amount of EPS, whereas the more extreme treatments like thermal exposure or addition of NaOH cause cell lysis, thereby leading to extracts containing both intra- and extracellular materials. Only those methods that are based on complexing agents to destabilize the EPS matrix avoid cell lysis while maintaining a relatively high yield. A complete removal of EPS would include the capsule that is tightly bound to the cell. It is generally assumed that this will invariably lead to a certain degree of cell lysis. For this reason the definition of extractable EPS must be an operational one. However, by taking the limitations of different extraction procedures into account, it is possible to choose a method that will allow an estimation of the influence of EPS on the properties of activated sludge and biofilms.

One of the most important questions when comparing methods for the extraction of EPS is the ability to control efficiency by knowing the theoretical yield of EPS. Our own attempts to determine EPS spectrophotometrically by using selective stains such as Calcofluor White have so far been unsuccessful because the same functional groups are present in EPS and on the cell surface (Spaeth, 1998). Figueroa and Silverstein (1989) developed a method that allowed the quantification of extracellular anionic polysaccharides and nucleic acids in activated sludge by in situ adsorption of the dye Ruthenium Red. The assumption is that the dye will bind only to EPS constituents and not to the cell envelope. The method has been applied to aerobic (Figueroa and Silverstein, 1989) and anaerobic sludge (Poxon and Darby, 1997). In the latter study 18–20% of the sludge total solids were reported to be EPS when expressed as equivalents of sodium alginate. The procedure is based on spectrophotometric determination of the nonadsorbed fraction of ruthenium red in the supernatant after centrifugation of exposed sludge. This does not allow an estimation of the composition of extracellular biopolymers. Interestingly, 18–40% of the measured EPS was found in the unwashed sludge. This would indicate that the common practice of first washing the sludge to separate loosely bound soluble EPS can account for a considerable fraction of the exopolysaccharides found in situ (the washing step is usually performed to remove recently adsorbed material). Whether this loosely bound material does in fact play a role in the properties attributed to EPS remains to be investigated. It should be emphasized that it is unclear whether any amount of Ruthenium Red also adheres to cell surfaces. Because ionized carboxyl groups on acidic polysaccharides are a likely binding site for this dye, the possibility cannot be ruled out that proteins also bind the dye.

QUANTIFICATION

The final step in the investigation of extracellular biopolymers is their quantification. Depending on the objectives chosen, biopolymers can be quantified on the basis of summaric parameters or by single substance analyses. One approach consists of gravimetric measurements that make use of the insolubility of biopolymers in organic solvents such as acetone or ethanol (Gehr and Henry, 1983; Karapaniagotis et al., 1989; Morgan et al., 1990). The results are expressed as acetone- or ethanol-insoluble material. Other summaric parameters like TOC can also be used to determine polymer concentrations in the EPS fraction (Morgan et al., 1990). By using simple measurements one can evaluate the efficiency of EPS separation or compare EPS contents of different sludge or biofilm samples. They do not allow an analysis of the composition or structure of EPS.

To learn more about the individual components of extracted EPS classical biochemical methods can be used for the analysis of individual substances or groups of substances. Care must be taken when determining proteins and other polymers in the resulting EPS extract to establish and validate methodological procedures that are affected by matrix effects and incompatibilities with constituents of the extraction buffers used. Table 2 lists the relative efficiency of EPS extraction of three methods based on the quantification of biopolymers. The comparison was performed by one laboratory. It is important to

TABLE 2. Comparison of EPS Extraction Methods as a Function of the Amount of Extracted Extracellular Biopolymer. All Data Are Expressed as mg/g Total Solids. Parentheses Indicate the Percentage of Total Biopolymer Content Found in the EPS (adapted from Spaeth et al., submitted).

	Protein	Carbohydrate	Uronic acid
Ion exchange resin	75.9 (18.6)	19.2 (4.9)	1.5 (1.9)
EDTA	26.1 (6.4)	2.5 (0.7)	0.5 (0.6)
Crown ether	21.3 (5.2)	14.5 (3.8)	0.9 (1.1)

note, however, that there are no standardized procedures for the determination of EPS components such as proteins or carbohydrates. Some workers have used the determination of total organic nitrogen according to Kjeldahl, using a conversion factor for the calculation of protein content based on the assumption that proteins contain 16.5% nitrogen (Raunkjaer et al., 1994). This assumption can be questioned because other compounds including amino acids, urea and humic acids are also counted by the Kjeldahl procedure. The most common methods for the determination of proteins are based on the procedures of Lowry (Lowry et al., 1951) and Bradford (1976). The Lowry method is based on the reaction of peptides with copper sulfate and the Folin Ciocalteau reagent. Usually, bovine serum albumin is used as the reference protein. Unfortunately, this procedure is prone to interferences by a variety of substances (Box, 1983). There have been attempts to quantify these effects by modifying the original method. For example, it was suggested to include an anionic detergent, sodium dodecyl sulfate (SDS), to increase the solubility of lipoproteins. In this way the interference of lipids was minimized with the unwelcome side effect of aiding in the solubilization and denaturation of membrane proteins and proteolipids. The presence of humic acids also interfered with the determination of proteins according to Lowry (Lowry et al., 1951). Their absorbance can be determined separately by a modification of the original procedure (Frølund et al., 1995). Bradford (1976) used the dye Coomassie Brilliant Blue G-2550 for the spectrophotometric determination of proteins. This method appears to be less susceptible to interfering substances, although this has not been studied to the extent of the investigations performed according to the Lowry method (Lowry et al., 1951). A disadvantage of this method is its differential sensitivity toward proteins, which can result in imprecise measurements (Raunkjaer et al., 1994). The different methods are compared in Table 3.

Carbohydrate determination is usually performed according to the anthron or phenol-sulfuric acid method (Dreywood, 1946; Dubois et al., 1956). Both methods show a high degree of specificity toward carbohydrates and are not affected by substances commonly found in wastewater. Nonetheless, hexoses and pentoses give different intensities of the resulting color complexes when using the anthron method, with hexoses resulting in more intense colors than pentoses (Koehler, 1952). Because glucose is usually the reference substance in the determination of carbohydrates, this can lead to errors in the presence of pentoses in the wastewater sample. Care must be taken also in the case of the glutaraldehyde extraction method (Azeredo et al., 1998) because glutaraldehyde may affect the analysis of carbohydrates by the phenol-sulfuric acid method as has been shown for formaldehyde (Underwood et al., 1995).

EPS may also be characterized by pyrolysis-mass spectrometry combined with multivariate statistical techniques. This allowed the detection of growth phase-related differences in EPS composition of aquatic bacteria (Ford et al.,

TABLE 3. Comparison of Different Methods for the
Determination of Carbohydrates and Protein in
Activated Sludge and EPS. All Data Are Expressed
as Mean ± Standard Deviation in mg/g VS.

Method of Analysis	Activated Sludge	EPS
Carbohydrates		
Phenol/sulfuric acid	219 ± 13	47 ± 7
Anthron	196 ± 10	41 ± 3
Proteins		
Bradford	NA	73 ± 6
Lowry	500 ± 18	352 ± 8
Lowry[a]	380 ± 11	212 ± 8
N-content	362 ± 36	NA

[a]Modified Lowry method.
NA = not analyzed.
Reprinted from *Water Research*, Vol. 30, Frølund et al., Extraction of extracellular polymers from activated sludge using a cation exchange resin, p. 1752, © 1996, with permission from Elsevier Science.

1991). The method has not yet been applied to the analysis of EPS isolated from wastewater.

CELL LYSIS

The separation of EPS without destruction of cells is an important prerequisite of any useful extraction method to avoid contamination of extracellular material with intracellular polymers or enzymes. Most methods that have been reported use rather extreme conditions that inevitably lead to a certain degree of cell lysis (Brown and Lester, 1980). Different methods can be used to assay for the extent of cell lysis that has occurred during the extraction of EPS. Intracellular enzymes like glucose-6-phosphate dehydrogenase (G6PDH) have been used (Frølund et al., 1996; Platt and Geesey, 1985) by determining specific activity according to Lessie and Van der Wyk (1972). However, it is not obvious from this method whether the enzyme has been released because of the rough extraction procedure or whether it leaked out of cells prior to extraction, due to autolysis. Another indicator compound for cell lysis is ribose as a constituent of nucleic acids, coenzyme A, ATP, FAD and NAD (Azeredo et al., 1998). A different approach is the determination of total cell numbers by applying nucleic acid stains such as acridine orange or 4',6-diamidino-2-phenylindole (DAPI) and redox dyes like CTC, which allow the microscopic determination of intact cells (Frølund et al., 1996). Recently, a new proprietary stain (Live/Dead by Molecular Probes, Oregon) that is claimed to distinguish between intact cells and cells with damaged cell envelopes has been applied successfully to biofilms (Neu and Lawrence, 1997).

Other workers have judged the extent of cell lysis based on the presence of DNA or protein in the EPS extract (Brown and Lester, 1980; Gehr and Henry, 1983). This method did not prove satisfactory because the above polymers have also been found as constituents of EPS that may have been accumulating from the surrounding water or as a result of lysis of cells (Rudd et al., 1983; Urbain et al., 1993). For example, about 40% of the total DNA content in activated sludge was found in EPS extracts (Nishikawa and Kuriyama, 1968). By using the cation exchange method to extract EPS from activated sludge (Frølund et al., 1996) DNA was found both in the bound EPS fraction and in the soluble fraction (Palmgren and Nielsen, 1996). Nevertheless, cell lysis can be followed during the development of an extraction procedure by observing the increase in DNA content while varying the length of extraction. In conclusion, any efficient EPS extraction procedure will result in a certain if minimal degree of contamination with intracellular material. The extent of tolerable lysis will determine which extraction procedure is best suited for a particular purpose.

APPLICATIONS

The range of applications of EPS extraction is manifold. Beginning in the 1950s EPS produced by axenic cultures were isolated for serological purposes (Dudman and Wilkinson, 1956; Davies, 1955). The investigations included an analysis of the changes in the formation and composition of EPS as a function of varying culture conditions. EPS of numerous isolates from different environments have been characterized including bacteria from activated sludge, water, soil and sediment (Ford et al., 1991). These studies have shown that EPS are highly diverse with both growth conditions and growth phase having a direct influence on the composition and quantity of extracellular biopolymers. There have been attempts to establish a link between structure and composition of EPS and the properties of the biological matrix being studied (Uhlinger and White, 1983; Sutherland, 1972). These investigations have prompted similar studies on EPS in biological wastewater treatment to determine if EPS can also influence the properties of activated sludge. The most important applications are flocculation, dewatering ability, sorption behavior, and biological activity of activated sludge.

The chemical composition of EPS affects the surface properties of the floc and thus the physical properties of activated sludge. A comparison of activated sludge before and after EPS extraction by centrifugation showed a reduction in the filterability and viscosity of the sludge after extraction (Sanin and Vesilind, 1994). Charged functional groups in EPS play an important role in flocculation and settleability of sludge (Pavoni et al., 1972; Steiner et al., 1976; Novak and Haugan, 1981). There was a correlation between EPS composition and surface charge of flocs when both aerobic and anaerobic sludge was analyzed (Morgan et al., 1990). Activated sludge had a protein:carbohydrate ratio of less than one, whereas the ratio was 3:1 in the case of anaerobic sludge. These results were later substantiated by other laboratories (Table 4). It should be noted, however, that the ratio of individual components within the EPS strongly depends on the experimental extraction procedure used and the type of sludge. For example, the protein:carbohydrate ratio was a function of extraction length and ranged from 3.9 to 5.1 in the case of an ion exchange-based extraction method (Frølund et al., 1996). Protein was the predominant exopolymer in activated sludge which had been adapted to a growth medium containing bovine serum albumin (Vallom and McLoughlin, 1984). These examples demonstrate that any direct comparison of individual studies pertaining to the extraction of EPS from a biological matrix is problematic in view of the variety of protocols used for extraction and analysis of individual components of the EPS.

EPS extraction has also been used to study the sorption behavior of biofilms and activated sludge toward metals. Different laboratories have shown that EPS that have been extracted from pure cultures can complex and accumulate metal ions (Mittelman and Geesey, 1985; Chen et al., 1995). Because EPS contain anionic groups such as those of uronic acids, it has generally been assumed that cations bind by an ion exchange mechanism. It was shown that EPS originating from different organisms display differing stability constants for complexes with metal ions, indicating the existence of distinct affinities for metals of individual extracellular biopolymers (Brown and Lester, 1982).

Another possibility for the analysis of metal sorption to biopolymers consists of separating EPS from the exposed biomass; only few methods are avail-

TABLE 4. Comparison of Protein: Carbohydrate Ratios in EPS.

Sludge Type	Protein: Carbohydrate	Reference
Activated	0.16	Horan and Eccles, 1986
Activated	0.56	Forster and Clarke, 1983
Activated	0.70	Morgan et al., 1990
Digested	3.0–3.6	Karapanagiotis et al., 1989
Digested	5.1	Forster, 1982
Digested		Ehlinger et al., 1987
(a) Glucose fed	0.61	
(b) VFA fed	2.31	
Various digesting sludges	1.1–2.8	Morgan et al., 1990

able for this approach. Methods that result in increased cell lysis (alkaline treatment) or a small yield of EPS (centrifugation) can only give a general indication of metal distribution. Other more effective techniques like extraction with EDTA or an ion exchange resin result in binding of metal by the extracting agent itself and thus influence the equilibrium within the biomass. Shen et al. (1993) used thermal extraction at 50°C to locate trace metals in the EPS of anaerobic granula. The above-described methodological inadequacies regarding limited yield and increased cell lysis apply. A new method for the quantification of metals in EPS is the use of a crown ether that selectively binds alkaline and alkaline earth metals (Figure 1) (Wuertz et al., submitted). The crown ether itself does not bind heavy metals, and the effectiveness of the extraction procedure is comparable with that of analogous methods including EDTA and cation exchange resin (Table 2). The method allows the direct determination of metals in EPS. Based on this method it was shown that in a biofilm exposed to cadmium and zinc, about 80% of the metal was bound by the cellular fraction (Figure 2). This result was surprising because EPS contain a number of possible binding sites (Spaeth et al., 1998). On the other hand, analyzing for lipophilic organic substances, e.g., benzene-toluene-xylene (BTX), more than 60% of the total content was found in the EPS (Figure 3). This was unexpected because of the generally hydrophilic character of the EPS.

In the treatment of wastewater, not only the sorption sites are of relevance but also microbial activity. The determination of enzymatic activity allows an assessment of biomass capacity for the degradation of organic contaminants. It was shown that a substantial amount of hydrolytic enzymes in activated sludge is bound to EPS (Frølund et al., 1995). The bioavailability of refractory and particulate substances is not only mediated by hydrolytic enzymes but also by extracellular redox activities that have been shown to increase in the presence of recalcitrant molecules (Wuertz et al., 1998). The separation of

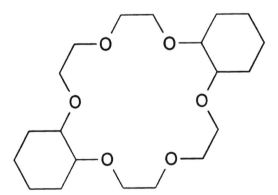

Figure 1 Structure of dicyclohexyl-18-crown-6.

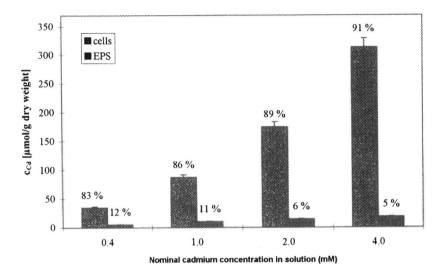

Figure 2 Distribution of cadmium in biofilms after EPS extraction with crown ether. Histograms are means (± SE) of triplicate determinations. The recovery of cadmium was about 95% (adapted from Spaeth et al., 1998).

EPS with ion exchange resin (Frølund et al., 1996) and subsequent addition of XTT (see Nomenclature), which is reduced to a colored water-soluble formazan salt, lead to an easy assay for the determination of extracellular redox enzymes (Griebe et al., 1997).

All of the above methods for the extraction of EPS should be weighed against their effectiveness in extracting biopolymers. Unfortunately, no satisfactory method that would allow an objective quantification of the amount of

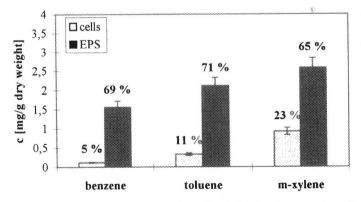

Figure 3 Distribution of benzene, toluene and m-xylene in biofilm after extraction with crown ether. Histograms are means (± SE) of triplicate determinations. The recovery of BTX was about 80% (adapted from Spaeth et al., 1998).

EPS present in a sample is available at present. The in situ method involving the stain Ruthenium Red (Figuera and Silverstein, 1989) is acceptable only under the premise that the compound is specific for EPS. The most recent attempt (Jahn and Nielsen, 1998; Münch and Pollard, 1997) by invoking cell numbers and cell volumes is based on approximation and microscopic analysis. The conversion factors for protein content and TOC introduce several unknowns that can slant all subsequent calculations and interpretations regarding the abundance and role of EPS in a sample. Nevertheless, this method represents the only approach known to the investigators that can give an approximate quantitative idea of the EPS present in an undisturbed sample.

Despite this lack of methodology, it must be stated that the more quantitative methods have been tested relative to one another, and they are suitable for an improved understanding of the composition and function of EPS and their influence on the properties of biofilm and activated sludge. The application of other tools such as NMR and FTIR spectroscopy to the noninvasive study of EPS may further aid in understanding the exact role of EPS and their relationship with the physiology of the microorganisms involved.

ACKNOWLEDGEMENTS

Our own research was supported by a grant to Stefan Wuertz, grant number Wu 268/2-1, from the German Research Foundation (DFG).

NOMENCLATURE

ATP:	Adenosine triphosphate
BSA:	Bovine serum albumin
BTX:	Benzene-toluene-xylene
CLSM:	Confocal laser scanning microscopy
CTC:	5-Cyano-2,3-di-4-tolyl-tetrazolium chloride
DAPI:	4',6-Diamidino-2-phenylindole
DNA:	Desoxyribonucleic acid
EDTA:	Ethylenediaminetetraacetic acid
EPS:	Extracellular polymeric substances
FAD:	Flavine adenine dinucleotide
FTIR:	Fourier transform infrared spectroscopy
NAD:	Nicotinamide adenine dinucleotide
NMR:	Nuclear magnetic resonance spectroscopy
SDS:	Sodium dodecyl sulfate
SEM:	Scanning electron microscopy
TEM:	Transmission electron microscopy

TOC: Total organic carbon
XTT: 3'-{1-[(phenylamino-)carbonyl]-3,4-tetrazolium}-bis (4-methoxy-6-nitro) benzene-sulfonic acid hydrate

REFERENCES

Azeredo, J., R. Oliveira and V. Lazarova. 1998. A new method for extraction of exopolymers from activated sludges. *Water Sci. Technol.* 37, 367–370.

Box, J.D. 1983. Investigation of the Folin-Ciocalteau phenol reagent for the determination of polyphenolic substances in natural waters. *Water Res.* 17, 511–525.

Bradford, M.M. 1976. A rapid and sensitive method for the quantitation of microgram quantities of protein utilising the principle of protein-dye binding. *Anal. Biochem.* 72, 248–254.

Brisou, J.F. 1995. *Biofilms: methods for enzymatic release of microorganisms.* CRC Press, Boca Raton, FL.

Brown, M.J. and J.N. Lester. 1980. Comparison of bacterial extracellular polymer extraction methods. *Appl. Environ. Microbiol.* 40, 179–185.

Brown, M.J. and J.N. Lester. 1982. Role of bacterial extracellular polymers in metal uptake in pure bacterial culture and activated sludge-I: effects of metal concentration. *Water Res.* 16, 1539–1548.

Bruus, J.H., P.H. Nielsen, and K. Keiding. 1992. On the stability of activated sludge flocs with implications to dewatering. *Water Res.* 26, 1597–1604.

Chen, J.H., L.W. Lion and W.C.S.M.L. Ghiorse. 1995. Mobilization of adsorbed cadmium and lead in aquifer material by bacterial extracellular polymers. *Water Res.* 29, 421–430.

Costerton, J.W. and R.T. Irvin. 1981. The bacterial glycocalyx in nature and disease. *Ann. Rev. Microbiol.* 35, 299–324.

Davies, D.A.L. 1955. The specific polysaccharides of some gram-negative bacteria. *J. Biochem.* 59, 696–704.

Del Gallo, M., M. Negi and C.A. Neyra. 1989. Calcofluor- and lectin-binding exocellular polysaccharides of *Azospirillium brasilense* and *Azospirillium lipoferum.* *J. Bacteriol.* 171, 3504–3510.

Dreywood, R. 1946. Qualitative test for carbohydrate material. *Ind. Eng. Chem.* 18, 499.

Dubois, M., K.A. Gilles, J.K. Hamilton, P.A. Rebers and F. Smith. 1956. Colorimetric method for determination of sugars and related substances. *Anal. Chem.* 28, 350–355.

Dudman, W.F. and J.F. Wilkinson. 1956. The composition of the extracellular polysaccharides of *Aerobacter-Klebsiella* strains. *J. Biochem.* 62, 289–295.

Ehlinger, F., J.M. Audic, D. Verrier and G.M. Fauq. 1987. The influence of carbon source on clogging in an anaerobic filter. *Water Sci. Technol.* 19, 261–273.

Figueroa, L.A. and J.A. Silverstein. 1989. Ruthenium red adsorption method for measurement of extracellular polysaccharides in sludge flocs. *Biotechnol. Bioeng.* 33, 941–947.

Flemming, H.C. 1991. Biofilms as a particular form of microbial life. In: H.-C. Flemming and G.G. Geesey (eds.): *Biofouling and biocorrosion in industrial water systems.* Springer, Heidelberg, Berlin, 5–11.

Flemming, H.C. 1995. Sorption sites in biofilms. *Water Sci. Technol.* 32, 27–33.

Flemming, H.C., J. Schmitt and K.C. Marshall. 1996. Sorption properties of biofilms. In: W. Calmano and U. Förstner (eds.): *Environmental behaviour of sediments.* Lewis Publishers, Chelsea, MI, 115–157.

Ford, T., E. Sacco, J. Black, T. Kelley, R. Goodacre, R.C.W. Berkeley and R. Mitchell. 1991. Characterization of exopolymers of aquatic bacteria by pyrolysis-mass spectrometry. *Appl. Environ. Microbiol.* 57, 1595–1601.

Forster, C.F. 1982. Sludge surfaces and their relation to the rheology of sewage sludge suspensions. *J. Chem. Technol. Biotechnol.* 32, 799–807.

Forster, C.F. and A.R. Clarke. 1983. The production of polymer from activated sludge by ethanolic extraction and its relation to plant operation. *Water Pollut. Control* 82, 430–433.

Frølund, B., T. Griebe and P.H. Nielsen. 1995. Enzymatic activity in the activated-sludge floc matrix. *Appl. Microbiol. Biotechnol.* 43, 755–761.

Frølund, B., R. Palmgren, K. Keiding and P.H. Nielsen. 1996. Extraction of extracellular polymers from activated sludge using a cation exchange resin. *Water Res.* 30, 1749–1758.

Geesey, G.G. and L. Jang. 1989. Interactions between metal ions and capsular polymers. In: T.J. Beveridge and R.J. Doyle (eds.): *Metal ions and bacteria.* John Wiley, New York, 325–357.

Gehr, R. and J.G. Henry. 1983. Removal of extracellular material, techniques and pitfalls. *Water Res.* 17, 1743–1748.

Goodwin, J.A.S. and C.F. Forster. 1985. A further examination into the composition of activated sludge surfaces in relation to their settlement characteristics. *Water Res.* 19, 527–533.

Griebe, T., G. Schaule and S. Wuertz. 1997. Determination of microbial respiratory and redox activity in activated sludge. *J. Ind. Microbiol. Biotechnol.* 19, 118–122.

Horan, N.J. and C.R. Eccles. 1986. Purification and characterisation of extracellular polysaccharide from activated sludges. *Water Res.* 20, 1427–1432.

Hsieh, K.M., G.A. Murgel, L.W. Lion and M.L. Shuler. 1994. Interactions of microbial biofilms with toxic trace metals. 1. Observation and modeling of cell growth, attachment, and production of extracellular polymer. *Biotechnol. Bioeng.* 44, 219–231.

Jahn, A. and P.H. Nielsen. 1998. Cell biomass and exopolymer composition in sewer biofilms. *Water Sci. Technol.* 37, 17–24.

Karapanagiotis, N.K., T. Rudd, R.M. Sterritt and J.N. Lester. 1989. Extraction and characterisation of extracellular polymers in digested sewage sludge. *J. Chem. Tech. Biotechnol.* 44, 107–120.

Koehler, L.H. 1952. Differentiation of carbohydrates by anthrone reaction rate and colour intensity. *Anal. Chem.* 24, 1576–1579.

Lessie, T.J. and J.C. Vander Wyk. 1972. Multiple forms of *Pseudomonas multivorans* glucose-6-phosphate and 7-phosphogluconate dehydrogenases: differences in size, pyridine specificity and susceptibility to inhibition by adenosine 5′-triphosphate. *J. Bacteriol.* 110, 1107–1117.

Lowry, O.H., N.J. Rosebrough, A.L. Farr and R.J. Randall. 1951. Protein measurements with the folin phenol reagent. *J. Biol. Chem.* 193, 265–275.

Mittelman, M.W. and G.G. Geesey. 1985. Copper-binding characteristics of exopolymers from a freshwater-sediment bacterium. *Appl. Environ. Microbiol.* 49, 846–851.

Morgan, J.W., C.F. Forster and L. Evison. 1990. A comparative study of the nature of biopolymers extracted from anaerobic and activated sludges. *Water Res.* 24, 743–750.

Münch, E. and P.C. Pollard. 1997. Measuring bacterial biomass-COD in wastewater containing particulate matter. *Water Res.* 31, 2550–2556.

Neu, T.R. 1996. Significance of bacterial surface-active compounds in interaction of bacteria with interfaces. *Microbiol. Rev.* 60, 151–166.

Neu, T.R. and J.R. Lawrence. 1997. Development and structure of microbial biofilms in river water studied by confocal laser scanning microscopy. *FEMS Microbiol. Ecol.* 24, 11–25.

Nielsen, P.H. and A. Jahn. 1999. Extraction of EPS. In: J. Wingender, T. Neu and H.C. Flemming (eds.): *Microbial extracellular polymer substances.* Springer Verlag, Heidelberg, 49–69.

Nielsen, P.H., A. Jahn and R. Palmgren. 1997. Conceptual model for production and composition of exopolymers in biofilms. *Water Sci. Technol.* 36, 11–19.

Nishikawa, S. and M. Kuriyama. 1968. Nucleic acid as a component of mucilage in activated sludge. *Water Res.* 2, 811–812.

Nivens, D.E., R.J. Palmer, Jr. and D.C. White. 1995. Continuous nondestructive monitoring of microbial biofilms: a review of analytical techniques. *J. Ind. Microbiol.* 15, 263–276.

Novak, J.T. and B.E. Haugan. 1981. Polymer extraction from the activated sludge. *J. WPCF* 53, 1420–1424.

Palmgren, R. and P.H. Nielsen. 1996. Accumulation of DNA in the exopolymeric matrix of activated sludge and bacterial cultures. *Water Sci. Technol.* 34, 233–240.

Pavoni, J.L., M.W. Tenney and W.F. Echelberger, Jr. 1972. Bacterial exocellular polymers and biological flocculation. *J. WPCF* 44, 414–431.

Pedersen, K. 1990. Biofilm development on stainless steel and PVC surfaces in drinking water. *Water Res.* 24, 239–243.

Platt, R.M. and G.G. Geesey. 1985. Isolation and partial chemical analysis of firmly bound exopolysaccharide from adherent cells of a freshwater sediment bacterium. *Can. J. Microbiol.* 31, 675–680.

Poxon, T.L. and J.L. Darby. 1997. Extracellular polyanions in digested sludge: measurement and relationship to sludge dewaterability. *Water Res.* 31, 749–758.

Raunkjaer, K., T. Hvitved-Jacobsen and P.H. Nielsen. 1994. Measurement of pools of protein, carbohydrate and lipid in domestic wastewater. *Water Res.* 28, 251–262.

Rudd, T., R.M. Sterritt and J.N. Lester. 1983. Extraction of extracellular polymers from activated sludge. *Biotechnol. Lett.* 5, 327–332.

Sanin, F.D. and P.A. Vesilind. 1994. Effect of centrifugation on the removal of extracellular polymers and physical properties of activated sludge. *Water Sci. Technol.* 30, 117–127.

Sato, T. and Y. Ose. 1980. Floc-forming substances extracted from activated sludge by sodium hydroxide solution. *Water Res.* 14, 333–338.

Schmitt, J., D. Nivens, D.C. White and H.C. Flemming. 1995. Changes of biofilm properties in response to sorbed substances—an FTIR-ATR study. *Water Sci. Technol.* 32, 149–155.

Shen, C.F., N. Kosariac and R. Blaszczyk. 1993. The effect of selected heavy metals (Ni, Co and Fe) on anaerobic granules and their extracellular polymeric substance (EPS). *Water Res.* 27, 25–33.

Spaeth, R. 1998. Role of extracellular polymeric substances from biofilms and activated sludge in sorption of contaminants. Ph.D. thesis, Technical University of Munich (in German).

Spaeth, R., H.C. Flemming and S. Wuertz. 1998. Sorption properties of biofilm. *Water Sci. Technol.* 37, 207–210.

Steiner, A.E., D.A. McLaren and C.F. Forster. 1976. The nature of activated sludge flocs. *Water Res.* 10, 25–30.

Stillmark, H. 1888. Über Rizin, ein giftiges Ferment aus dem Samen von Rizinus communis. Ph.D. thesis, University of Dorport (in German).

Surman, S.B., J.T. Walker, D.T. Goddard, L.H.G. Morton, C.W. Keevil, W. Weaver, A. Skinner, K. Hanson, D. Caldwell and J. Kurtz. 1996. Comparison of microscope techniques for the examination of biofilms. *J. Microbiol. Methods* 25, 57–70.

Sutherland, I.W. 1972. Bacterial exopolysaccharides. *Adv. Microbiol. Physiol.* 8, 143–213.

Uhlinger, D.J. and D.C. White. 1983. Relationship between physiological status and formation of extracellular polysaccharide glycocalyx in *Pseudomonas atlantica*. *Appl. Environ. Microbiol.* 45, 64–70.

Underwood, G.J.C., D.M. Paterson and R.J. Parkes. 1995. The measurement of microbial carbohydrate exopolymers from intertidal sediments. *Limnol. Oceanogr.* 40, 1243–1253.

Urbain, V., J.C. Block. and J. Manem. 1993. Bioflocculation in activated sludge: an analytical approach. *Water Res.* 27, 829–838.

Vallom, J.K. and A.J. McLoughlin. 1984. Lysis as a factor in sludge flocculation. *Water Res.* 18, 1523–1528.

Wolfaardt, G.M., J.R. Lawrence, R.D. Robarts and D.E. Caldwell. 1995. Bioaccumulation of the herbicide Diclofop in extracellular polymers and its utilization by a biofilm community during starvation. *Appl. Environ. Microbiol.* 61, 152–158.

Wuertz, S., P. Pfleiderer, K. Kriebitzsch, R. Spaeth, T. Griebe, D. Coello-Orviedo, P.A. Wilderer and H.C. Flemming. 1998. Extracellular redox activity in activated sludge. *Water Sci. Technol.* 37, 379–384.

Wuertz, S., R. Spaeth, A. Hinderberger, T. Griebe and H.-C. Flemming. Submitted for publication. A new method for extraction of extracellular polymers from biofilm and activated sludge based on dicyclohexyl-18-crown-6.

CHAPTER 5

Cell Biomass Determination in Sewer Biofilms and a Monospecies *Pseudomonas Putida* Biofilm

ANDREAS JAHN
PER HALKJÆR NIELSEN

INTRODUCTION

In biofilms, cells are embedded in a matrix of extracellular polymeric substances (EPS). The relative quantity of cell biomass and EPS varies largely in different biofilms. The cell biomass can account for 10–90% of the organic matter (Nielsen et al., 1997). The amount of EPS is important for many biofilm properties, e.g., strength, volumetric specific activity and effective diffusion coefficient in the biofilm. If these parameters are to be determined precisely in biofilm experiments, it is important to be able to divide total biofilm biomass into cell biomass and an EPS fraction.

A crucial point is the definition of the term "cell biomass." Here all macromolecules outside the cell not covalently attached to membrane structures of the cells will be called "extracellular." Slimes, capsules, S-layers and sheaths are defined as extracellular material in the biofilm in addition to lysis products and adsorbed matter. Note that in the microbial literature, cell biomass was defined as *cell biomass without storage material* (Pirt, 1982). The definition of cell biomass in this article also includes storage material inside the cells.

Different methods exist to determine cell biomass (Figure 1) (Fry, 1988). Direct methods are based on direct enumeration of bacterial cells. Usually, a direct staining of cells with a fluorescent dye such as 4′,6-diamidino-2-phenylindole (DAPI) or acridine orange (AO) at DNA-specific wavelengths is used. Cells are then detected in a fluorescence microscope. The total number of bacteria or, after determination of the mean size value, the total biovolume is converted into cell biomass. The common unit is total organic carbon (TOC), which can subsequently be related to TOC measured in the total sample. Another possibility of determining cell biomass is by extracting the EPS mater-

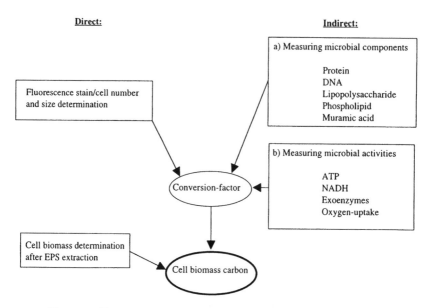

Figure 1 Different methods for direct and indirect cell biomass determination.

ial directly from the biofilm sample, which necessitates the existence of a quantitative extraction procedure for the EPS material.

Indirect methods for cell biomass determination can be a measurement of certain marker molecule or microbial activity assumed to be unique for cell biomass. Protein, DNA, phospholipids, lipopolysaccharide, muramic acids or metabolites ATP and NADH have been used as such indirect marker molecules. Other activity measurements use exoenzymatic activity or oxygen utilization for biomass determination (Lazarova and Manem, 1995). Indirect measurements, and in particular the activity measurements, are usually used to determine biomass and not only cell biomass. When using indirect methods, a conversion factor must also be used to convert the parameters measured into cell biomass.

This chapter presents results obtained by using different direct and indirect methods for determination of cell biomass and EPS in sewer biofilms and in a monospecies biofilm culture of *Pseudomonas putida*. The monospecies biofilm allowed us to perform experiments with biofilms grown under well-controlled conditions. In addition, cells growing in a monospecies biofilm could be compared directly to cells growing in suspension. The results are discussed in view of the particular problems biofilms cause for different methods of cell biomass determination.

DIRECT DETERMINATION

Three steps are involved when determining cell biomass directly with epi-fluorescence microscopy: determination of cell number requiring a complete homogenization of the sample, determination of the mean cell volume, and selection of a conversion factor from cell volume to cell biomass. With these parameters the total biovolume and total cell biomass can be determined.

HOMOGENIZATION OF THE SAMPLE

Samples were mixed and homogenized by using a glass homogenizer with a gap size of 30–40 μm. The standard deviation of the cell number determined in different sewer lines after staining with AO was typically about 20–50% of the measured mean value (Figure 2). The main difficulty in determining the cell number is to resolve the microcolony structure frequently found in biofilm systems. After homogenization, biofilm samples contained bacterial aggregates of variable sizes (up to 20 μm in diameter) non-homogeneously distributed in the sample. Different detergents were tried to obtain a better homogenization of biofilm bacteria from different sewer samples. After applying anionic sodium dodecyl sulfate (SDS) [0.01% (w/v)] (Figure 2) or the non-ionic detergents Triton-X-100 [0.1% (v/v)] or Nonident-P-40 [0.1% (v/v)] (data not shown), no change in the mean value of the cell count was observed. Large particles of bacterial aggregates were still detected in the sample. To our knowledge, no method is available for the dissolution of all the bacterial colonies to single bacteria. Attention should be paid to these microcolonies and they should be carefully monitored. The microcolonies are also a major problem for automatic analysis of samples with image analysis equipment.

CELL VOLUME DETERMINATION

The mean cell volume was determined for sewer biofilm (Figure 3). In this analysis, 50% of the cells were smaller than 0.18 μm^3. A large variation in the sizes was observed in different samples. In the example shown in Figure 3, at least two classes of cell volumes could be found: very small cells with a cell volume smaller than 0.1 μm^3 which made up 40% of the total sample. Another 40% of the cells determined were between 0.14 and 0.3 μm^3. In all cell volume determination of cells from biofilms, a large standard deviation of the mean cell volume was observed. It was typically the size of the mean value of the cell volume.

The reason for the variation observed is twofold. First, in natural systems such as sewer biofilms examined in this study, cells vary in size, depending on the species. Second, cells are known to obtain different sizes, depending

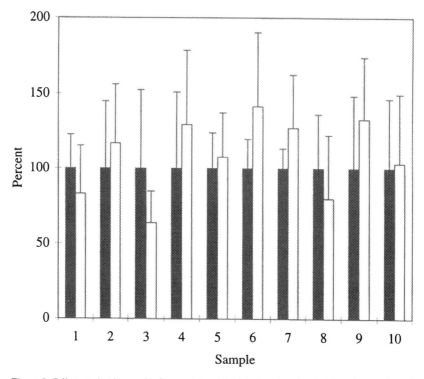

Figure 2 Cell counts in 10 sewer biofilm samples with (closed columns) and without (open columns) applying SDS in a concentration of 0.01% (w/v) during homogenization procedure. Cell number was determined by epifluorescence microscopy. One milliliter of homogenized biofilm sample was stained with 1 ml of acridine orange solution (0.1% in aq. dist) and counted on a Costar polycarbonate filter (0.2 μm) using a Leitz epifluorescence microscope with a 100 × objective.

on the nutrient supply (Mason and Egli, 1993). It is a well-known fact that bacteria become smaller and grow into different shapes when being starved of nutrients. In biofilms, substrate gradients are always present, and this results in variable cell sizes. Dependent on the type of biofilm examined, it may be appropriate to use different classes of cell volumes for biomass determination. The use of confocal laser microscopy is expected to give new technical possibilities for the correct total cell volume determination in biofilms.

DETERMINING CELL VOLUME/CELL BIOMASS CONVERSION FACTOR

The biovolume of a sample is usually converted into a TOC value for cell biomass determination. Different conversion factors are reported in the literature, and they vary at least by a factor of three. A mean cell volume/TOC

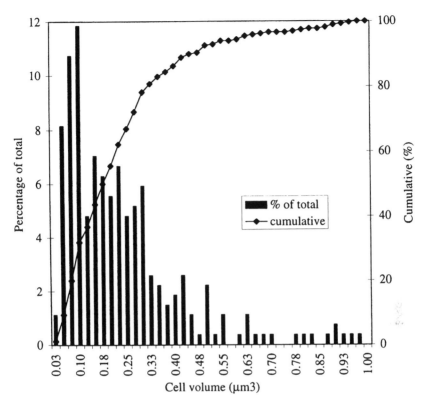

Figure 3 Cell volume distribution of a sewer biofilm. Cells were stained with acridine orange as described in Figure 2. The cell volume was determined by using a CCD-camera (Image-point) and IPLab software to measure the length and width of the bacteria. The cell volume was calculated considering bacteria straight-sided rods with hemispherical ends (Fry, 1988).

conversion factor of 308 fg/μm^3 was recommended (Fry, 1988), which may be appropriate for biofilm systems with unknown bacterial composition. When working with monospecies biofilms or defined mixed cultures, conversion factors can be determined directly from suspended growing cells. The question is, under which culture conditions should a reference cell be grown.

Cell morphology and TOC value per cell volume for a *P. putida* culture were determined in detail to study the importance of growth stages for a single species. Batch cultures in the exponential growth phase were compared with a culture after 4 days of starvation without any carbon source. Cells in exponential phase were rod-shaped and longer (1.68 μm in length) than the small coccoid-like cells 0.70 μm long after carbon starvation (Table 1). The results clearly show that the carbon content per cell volume depended on the physiological state of the cells with the highest values for cells in the starvation phase.

TABLE 1. Cell Size and Carbon Content Per Cell Volume of Exponential Growing and Starvation Phase Cells of *P. putida* After 4 Days of Carbon Starvation. SD of the Determinations ($\pm$) is Reported. Cells Were Incubated at 30°C at 150 rpm in a Medium Containing 10 g/l Citrate, 2 g/l $(NH_4)_2SO_4$, 6 g/l Na_2HPO_4, 3 g/l KH_2PO_4, 3 g/l NaCl, 1 mM $MgCl_2$, 0.4 mM $CaCl_2$, 0.01 mM $FeCl_3$, 2.5 mg/l Thiamine, pH 7.2. In the Exponential Phase, Cells were Harvested by Centrifugation (5000 · g, 10 min) and Incubated for Starvation on the Medium Without Carbon Source. Slimes Were Removed by Centrifugation (5000 · g, 10 min) and Total Organic Carbon (TOC) Was Determined After Acidification and Purging of the Inorganic Carbon With Persulfate Oxidation in an O.I. Analytical Model 700 Carbon Analyzer.

	Exponential Phase	Stationary Phase
Length (μm)	1.68 $\pm$ 0.56	0.70 $\pm$ 0.17
Width (μm)	0.67 $\pm$ 0.19	0.51 $\pm$ 0.10
Cell volume (μm^3)	0.51	0.11
Cells/ml ($\times$ 10^8)	3.7 $\pm$ 0.9	6.9 $\pm$ 0.6
TOC/cell volume (fg/μm^3)	250	560

Another possibility of determining cell biomass/TOC conversion factor is to grow cells in a chemostat under defined limitation. In a carbon-limited chemostat the cell dimension was compared with the cell dimensions obtained from batch studies. At a dilution rate of 0.23 h^{-1} the mean length of the carbon limited cells was 1.41 $\pm$ 0.40 μm and the mean width 0.66 $\pm$ 0.12 μm. This resulted in a mean cell volume of 0.41 μm^3, which corresponded to cells from the exponential phase in batch cultivation. In carbon-limited *P. putida* biofilms a mean value of 0.22 μm^3 was found (Table 2).

These results illustrate that conversion factors should be considered carefully. From the results reported here and based on the observation of others (Lee and Fuhrmann, 1987), it is known that the macromolecular content per

TABLE 2. Biofilm Parameters of a 120-hr Citrate Limited *P. putida* Biofilm. The Biofilms Were Run With $D = 1.8$ h^{-1} in RotoTorque Reactors (Characklis, 1990) From SINIS GmbH, Stuttgart, Germany. The Medium Consisted of 0.1 g/l Citrate, 0.2 g/l $(NH_4)_2SO_4$, 1.5 g/l Na_2HPO_4, 0.75 g/l KH_2PO_4, 1.3 g/l NaCl, 1 mM $MgCl_2$, 0.4 mM $CaCl_2$, 0.01 mM $FeCl_3$, 0.6 mg/l Thiamin, pH 7.2, 30°C.

TOC/cm^2 (μg)	137.8
Protein/cm^2 (μg)	124.1
DNA/cm^2 (μg)	3.1
Cells/cm^2 ($\times$10^8)	2.58
Mean cell volume (μm^3)	0.22

cell volume may vary at least by a factor of two, mainly because the smaller cells have a higher relative concentration of macromolecules. As discussed above, because of substrate gradients present in biofilms, the cells may have a non-homogeneous distribution of cell sizes. It may be appropriate to use conversion factors determined after substrate starvation in batch cultures, in particular for monoculture biofilms or defined mixed cultures.

EXTRACTION OF EXTRACELLULAR POLYMERS

Cell biomass in biofilm systems can also be determined directly after extraction of EPS material from the biofilms if the extraction procedure is quantitative. The fraction remaining after an extraction should correspond to the biomass. In suspended growth, cells may easily be separated from the slime macromolecules in the surrounding medium by centrifugation or filtration. Because of the adhesive properties of the molecules in the biofilm matrix this is

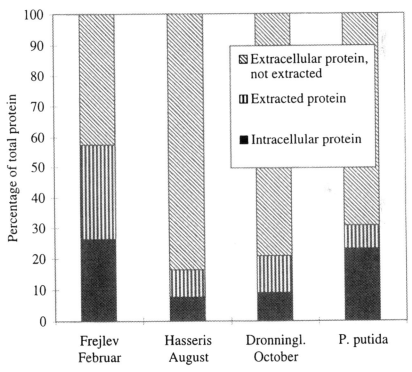

Figure 4 Distribution of protein in different biofilms. Cell protein fraction was calculated from the biovolume TOC by using a carbon/protein conversion factor of 0.9. The Dowex extractable fraction was extracted with Dowex. Protein was determined by using a modified Lowry method (Frølund et al., 1996).

TABLE 3. Dowex Extractable Fraction From Different Sewers From the Aalborg Area and a Carbon Limited Biofilm Culture of *P. putida*. The Extraction Efficiency is Shown in Brackets. Samples Were Extracted by Means of a Dowex Cation Exchange Resin (Jahn and Nielsen, 1995). *P. putida* Biofilms Were Run as Described for Table 2. Carbohydrate was Determined by Using the Anthrone Method (Raunkjaer et al., 1994). Uronic Acids Were Determined by the *m*-hydroxydiphenyl Sulfuric Acid Method (Kintner and van Burren, 1982). DNA was Determined by Using the DAPI Method (Brunk et al., 1979). Protein and Humic Substances Were Determined by the Lowry Method and Corrected for the Humic Substances. Humic Substances Were Measured With a Modified Lowry Procedure (Frølund et al., 1996).

	Protein (mg/l)	Humic Substances (mg/l)	Polysaccharide (mg/l)	Uronic Acids (mg/l)	DNA (mg/l)
Frejlev	1000 (31)	1281 (58)	60 (4)	26 (22)	30
Hasseris	1000 (9)	441 (39)	206 (8)	17 (14)	25
Dronningld.	1000 (11)	629 (8)	132 (4)	32 (19)	35
P. putida	1000 (7.3)	—	96 (13)	13 (15)	60 (22)

not so easy for biofilm cultures. An extraction by centrifugation is not enough for biofilm EPS (Platt et al., 1985). Because of the different nature of the forces keeping a biofilm together (at least electrostatic and hydrophobic interactions are involved), it cannot be expected to find one extraction method that would be valid for all EPS in the biofilm. Among the more promising methods for extraction of EPS from biofilms is a cation exchange mechanism using EDTA or Dowex (Jahn and Nielsen, 1995; Platt et al., 1985).

It is highly doubtful whether extractions can be performed quantitatively. This is illustrated with some results obtained from different sewer biofilms and a monoculture biofilm of *P. putida,* where the Dowex extraction method was tested. The amount of EPS protein extracted was calculated. Cell numbers and total cell volumes were determined in the biofilm samples. From the biovolume the total amount of carbon and protein in the cells were calculated. The conversion factor used for carbon was 308 $fgC/\mu m^3$. Protein typically makes 50% of the total carbon (Russel and Cook, 1995) and is a rather conservative amount of cell biomass. The distribution of protein in the biofilm was then calculated from a mass balance taking account of total protein, protein extracted by Dowex and the remaining EPS part (Figure 4). The results showed that 44–84% of the extracellular protein was still left in the different biofilms after Dowex extraction. The exact value will of course depend on the conversion factor used. However, even a two times higher conversion factor for cell carbon would result in a considerable amount of protein in the EPS. The results strongly suggested that the extraction of EPS protein from the biofilms investigated was not quantitative.

The extraction efficiencies varied largely for the different substances, but also within the same group. For example, humic substances that are expected only to occur outside the cells were extracted from sewer biofilms with 8–58% efficiency (Table 3). This shows that neither were humic substances quantitatively extracted nor was a common extraction efficiency for one group of substances from different biofilms. The Dowex extraction does not extract the EPS quantitatively. Depending on the properties of the macromolecules, a larger or smaller amount will be left with the cell fraction. This amount may vary from one biofilm to another, so a separation of extracellular polymers from the cell biomass is regarded as an inadequate method of determining cell biomass in biofilms.

INDIRECT METHODS

Protein and DNA have been used for biomass determination. Protein in particular seems to be valid in suspended culture systems because normally it will not accumulate as intracellular storage material and has a relatively constant value referring to the total carbon content of 50% of the cell biomass

(Russel and Cook, 1995). As shown above, a large part of the total biofilm protein seemed to be extracellular.

Different sewer biofilms and a pure culture of *P. putida* were analyzed for the EPS composition of the Dowex extractable fraction (DEPS) (Table 3). In all samples we found a highly heterogeneous matrix with respect to the DEPS composition. In sewers, protein and humic substances were the most pronounced fractions extracted, whereas in *P. putida,* biofilm protein constituted the main fraction of the exopolymers. DNA was found in the extracted fraction, too. In monospecies biofilms of *P. putida* 22% of the total DNA were found in the EPS material.

DNA content of cells is known to be dependent on the growth rate of the microorganism and will vary by a factor of four during the growth cycle (Mason and Egli, 1993). In addition, large plasmids may be present and also give variation in results. DNA was also isolated from the slimes from *P. putida* pure culture in a chemostat, from activated sludge (Palmgren and Nielsen, 1996) and sewer biofilms (Jahn and Nielsen, 1995). The uncertainty in determining cell biomass with DNA will depend on the growth state of bacteria present and on the molecules in the matrix. DNA bound to particles has a much higher stability than free DNA (Lorenz and Wackernagel, 1987) and in discussing degradation of DNA outside the cells, this particle binding has to be taken into account.

COMPARISON OF DIRECT AND INDIRECT CELL BIOMASS DETERMINATION IN A *P. PUTIDA* BIOFILM

TOC, protein, DNA, number of bacteria and the mean bacterial cell volume were determined in a carbon limited monoculture biofilm of *P. putida.* Cell biomass was calculated after cell count and biovolume determination with two different cell volume/cell carbon conversion factors from the exponential phase and from the stationary phase after starvation (Table 1). By way of comparison, both protein and DNA were used as marker molecules for indirect cell biomass determination. Cell biomass was calculated with DNA/TOC and protein/TOC factors determined in continuous cultivation of *P. putida* (Table 4).

When total protein was used as a measure of biomass, a value of almost 81% of total TOC was found to be cell biomass (Figure 5); and 7.4% of the protein could be extracted with Dowex. After correction of the extracted protein, 74% of TOC was calculated as cell biomass. In a similar way DNA was used as biomass marker. The calculated cell biomass corresponded to the values found by cell number/cell volume determination when a high cellular carbon content (560 fg/μm^3) was used for the calculation.

It can be concluded that the direct and indirect determination gave very different results. By direct analysis, 10–28% were cell biomass polymers. We know that a certain fraction of the protein (and DNA) is extracellular. For this

TABLE 4. Protein/Cell Biomass and DNA/Biomass Conversion Factors for *P. putida.* The Values Were Determined With Citrate Limited Steady-State Chemostat Cells at $D = 0.23$ h^{-1}. Medium As Described in Table 1.

Protein (μg) $\rightarrow$ cell biomass carbon (μg)	0.9
DNA (μg) $\rightarrow$ cell biomass carbon (μg)	16.6

reason, direct cell counts and biovolume determination seem to be the best way of determining cell biomass in biofilms because these methods do not depend on other parameters such as exopolymer production and cellular lysis.

It is known that the biofilm strength and properties depend on substrate conditions, shear and ionic composition. The biofilm strength may be correlated to the amount (Applegate and Bryers, 1991) and type of EPS (Nielsen et al., 1997). The EPS composition of the matrix was shown to interfere with the indirect biomass determination. It can also be anticipated that biofilms of different ages will have different amounts and types of EPS. On the basis of these considerations, it is extremely important to know what substances accumulate in the biofilm matrix when comparing different biofilms by means of indirect methods measuring microbial components.

The results presented for indirect determination of biomass were based on analysis of protein and DNA. Basically, the same considerations can be made for other indirect cell biomass determinations that are also reported in the literature. Lipopolysaccharides may accumulate in the biofilm matrix and thus lead to an incorrect determination of cell biomass in biofilm systems (Okabe et al., 1994). Extractable lipid phosphate was used for biomass determination in biofilm reactors and in sediments (Findlay et al., 1989; Hooijmans et al., 1995). Phospholipids are known to have a high turnover in sediments and can be extracted quantitatively (Findlay et al., 1989), making them potential candidates for cell biomass determination. However, for both lipopolysaccharides and phospholipids it is difficult to find a conversion factor for natural systems. Gram (+) bacteria do not have an outer membrane and will have a lower phospholipid content than Gram (−) bacteria (Gottschalk, 1986).

ATP and NADH may be the best molecules for an indirect determination of bacterial biomass. These molecules do not accumulate in the biofilm matrix. However, the ATP amount per cell as well as the NADH level depends on the growth state of the cell (Lazarova and Manem, 1995), so variations in the physiological state of the cells in the biofilm are important.

Exoenzymatic activity and oxygen utilization rates have also been used in biofilm systems. Exoenzymatic activity is hardly a proper measurement for cell biomass due to accumulation of enzymes in the EPS matrix. Such an accumulation has been described in the floc matrix of activated sludge (Frølund

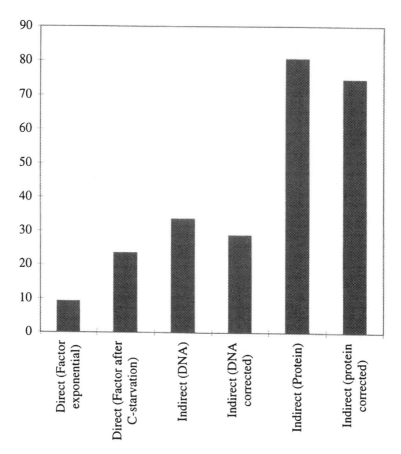

Figure 5 Example of calculation of cell biomass using different direct or indirect determination methods. Two different conversion factors (from exponential growth and after C-starvation) were used for direct cell biomass determination. For indirect measurements with DNA and protein, both the measured value and a value corrected for the Dowex extractable fraction are reported. Samples were measured in a *P. putida* monoculture biofilm after 4 days of growth. For growth conditions see Table 2.

et al., 1995). Oxygen utilization is problematic because of difficulties with the preincubation of the cells and a need for a conversion factor. In addition, it is not very sensitive compared with other methods (Lazarova and Manem, 1995)

CONCLUSIONS

The cell biomass in biofilms can be determined by direct and indirect methods. Direct cell biomass determination by cell count and cell volume measurements using fluorescence microscopy is considered the most reliable in most systems but has two difficulties: (1) determination of the correct number and size of bacteria and (2) finding a cell volume/cell biomass conversion factor. For environmental samples it is probably sufficient to use published factors for cell volume/cell biomass conversion. In defined culture systems, the bacterial species and the growth state affect the conversion factors and should be carefully considered. Chemostat grown cells can be difficult to use as reference for biofilm systems because the growth situation may not reflect the situation for most of the cells within a biofilm. The indirect methods for cell biomass determination (e.g., protein, DNA, phospholipids) have also the disadvantage that the conversion factor is usually not well known and it depends on a number of factors. Furthermore, a significant accumulation of macromolecules in the biofilm matrix can take place but is so far not well described. The direct cell biomass determination seems to be the best method especially for natural systems and should always be included in measurement of biofilm cell biomass and also when other indirect methods are used.

REFERENCES

Applegate, D.H. and J.D. Bryers. 1991. Effects of carbon and oxygen limitations and calcium concentrations on biofilm removal processes. *Biotechnol. Bioeng.* 37, 17–25.

Brunk, C.F., K.C. Jones and T.W. James. 1979. Assay for nanogram quantities of DNA in cellular homogenates. *Anal. Biochem.* 92, 497–500.

Characklis, W.G. 1990. Laboratory biofilm reactors. In: W.G. Characklis and K.C. Marshall (eds.): *Biofilms.* John Wiley & Sons, Inc., New York, 55–89.

Findlay, R.H., G.M. King and L. Watling. 1989. Efficacy of phospholipid analysis in determining microbial biomass in sediments. *Appl. Environ. Microbiol.* 55, 2888–2893.

Frølund, B., T. Griebe and P.H. Nielsen. 1995. Enzymatic activity in the activated-sludge floc matrix. *Appl. Microbiol. Biotechnol.* 43, 755–761.

Frølund, B., R. Palmgren, K. Keiding and P.H. Nielsen. 1996. Extraction of extracellular polymers from activated sludge using a cation exchange resin. *Water Res.* 30, 1749–1758.

Fry, J.C. 1988. Determination of biomass. In: B. Austin (ed.): *Methods in aquatic bacteriology.* John Wiley & Sons, Ltd., New York, 27–72.

Gottschalk, G. 1986. *Bacterial metabolism.* Springer Verlag, New York, Berlin.

Hooijmans, C.M., T.A. Abdin and G.J. Alaerts. 1995. Quantification of viable biomass in biofilm reactors by extractable lipid phosphate. *Appl. Microbiol. Biotechnol.* 43, 781–785.

Jahn, A. and P.H. Nielsen. 1995. Extraction of extracellular polymeric substances (EPS) from biofilms using a cation exchange resin. *Water Sci. Technol.* 32, 157–164.

Kintner, P.K. and J.P. van Burren. 1982. Carbohydrate interference and its correction in pectin analysis using *m*-hydroxydiphenyl method. *J. Food Sci.* 47, 756–760.

Lazarova, V. and J. Manem. 1995. Biofilm characterization and activity analysis in water and wastewater treatment. *Water Res.* 29, 2227–2245.

Lee, S. and J.D. Fuhrmann. 1987. Relationships between biovolume and biomass of naturally derived marine bacterioplankton. *Appl. Environm. Microbiol.* 53, 1298–1303.

Lorenz, M.G. and W. Wackernagel. 1987. Adsorption of DNA to sand and variable degradation rates of adsorbed DNA. *Appl. Environ. Microbiol.* 53, 2948–2952.

Mason, A. and T. Egli. 1993. Dynamics of microbial growth in the decelerating and stationary phase of batch culture. In: S. Kjelleberg (ed.): *Starvation in bacteria.* Plenum Press, New York, London, 81–102.

Nielsen, P.H., A. Jahn and R. Palmgren. 1997. Conceptual model for production and composition of exopolymers in biofilms. *Water Sci. Technol.* 36, 11–19.

Okabe, S., P.H. Nielsen, W.L. Jones and W.G. Characklis. 1994. Estimation of cellular and extracellular carbon contents in sulfate-reducing bacteria biofilms by lipopolysaccharide assay and epifluorescence microscopic technique. *Water Res.* 28, 2263–2266.

Palmgren, R. and P.H. Nielsen. 1996. Accumulation of DNA in the exopolymeric matrix of activated sludge and bacterial cultures. *Water Sci. Technol.* 34, 233–240.

Pirt, S.J. 1982. Maintenance energy: a general model for energy limited and energy sufficient growth. *Arch. Microbiol.* 133, 300–302.

Platt, R.M., G.G. Geesey, J.D. Davis and D.C. White. 1985. Isolation and partial chemical analysis of firmly bound exopolysaccharide from adherent cells of a freshwater bacterium. *Can. J. Microbiol.* 31, 675–680.

Raunkjær, K., T. Hvitved-Jacobsen and P.H. Nielsen. 1994. Measurement of pools of protein, carbohydrate and lipid in domestic wastewater. *Water Res.* 28, 251–262.

Russel, J.B. and G.M. Cook. 1995. Energetics and bacterial growth: balance on anabolic and catabolic reactions. *Microbiol. Rev.* 59, 48–62.

CHAPTER 6

Simultaneous Morphological and Population Analyses of Environmental Biofilms

JANINE A. FLOOD
NICHOLAS J. ASHBOLT
PETER J. BEATSON

PRINCIPLE OF THE METHOD

THE shortage of simple in situ techniques with which to study environmental biofilms has resulted in the present lack of quantitative data on the architecture and function of these microbial structures, even though they play vital roles in water quality, wastewater treatment and industrial biofouling. Thin sectioning is widely performed on many biological specimens to permit visualization of microscopic structure within macroscopic samples. This approach however, is generally applied to tissues that are easily handled, rather than delicate surface layers such as biofilm. Although cryosectioning methodologies have previously been applied to artificial biofilm, the technique had been shown to be unsuitable for many environmental biofilms because of the more heterogeneous composition of natural biofilms and their substrata (Huang et al., 1996).

This chapter outlines modifications and optimization procedures necessary for the successful application of cryosampling and cryosectioning to environmental biofilms. Difficulties arising from some sample types are indicated, and options to overcome these problems are presented and discussed. Subsequent application of some simple and complex staining procedures are described for environmental biofilm cryosections with examples shown of fluorescent in situ hybridization (FISH), microautoradiography and electron dispersive X-ray analysis (EDX). An overview of the cryosectioning method is given in Figure 1.

Specimen freezing facilitates the sampling of biofilm and minimizes handling damage, thus enabling the rapid and inexpensive collection of intact biofilm samples from a wide array of environmental substrata such as rocks, plants, pipelines and membrane filters. Frozen biofilm samples can be stored

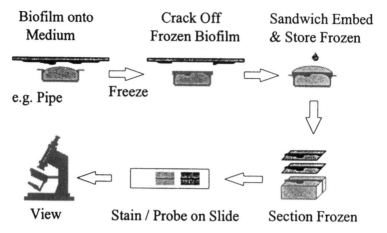

Figure 1 Diagram showing the cryosectioning processes for biofilm from unsectionable sub-strata. Samples are embedded and rapidly frozen to maintain sample integrity and then fractured from the frozen substratum, before being sandwich embedded for cryosectioning and subsequent staining and microscopy.

indefinitely and handled easily with a minimum of time, expense or equipment. Cryosectioning permits powerful biofilm analysis via the application of multiple histochemical and molecular techniques to a single sample while maintaining the in situ morphology. Characterization of environmental biofilms by cryosectioning and histochemical techniques will facilitate the collection of data on in situ morphology, population structure and physiological activity in actual biofilms from many varied fields of research and industry.

RANGE OF APPLICATIONS

The methods presented here are suitable for:

- wastewaters (domestic, industrial and agricultural; aerobic and anaerobic)
- clear waters (potable and cooling)
- natural environments (riverine and wetland)
- substrata such as rocks, pipes (cement, plastic, metal), sludge flocs, plants, membranes, filters and natural fibers

They are not suitable for soil or granular activated carbon (GAC).

METHODS

Various combinations of the procedures described below have been applied to a range of biofilm/substratum samples, establishing the suitability of this approach for most environmental biofilms. Samples tested include:

- wetland plants (stems and roots)
- stream pebbles
- polypropylene hollow-fiber microfilters (0.2 μm) processing pre-screened (200 μm) primary sewage
- reverse-osmosis membrane filters treating chlorinated, filtered sewage
- cement-lined steel pipes exhumed from a drinking water distribution system
- sludge granules and coconut husk from an anaerobic reactor processing abattoir wastewater
- polycarbonate and glass slides from laboratory reactors fed chloraminated drinking water
- matrices from various biofilm reactors, including vermiculite, gravel, plastics and granular activated carbon (GAC)

SAMPLING

When obtaining biofilm-covered specimens, it is generally advisable to handle the objects with great care and with as little movement as possible, to avoid dislodging any loosely attached components. In our experience, passage of wetland biofilms through the air/water interface did not result in any detectable alteration of the biofilm morphology, compared with samples that were kept wholly submerged until after embedding. This cannot be assumed to hold true for all biofilm types, however, and therefore where possible, such comparisons should be conducted for all new biofilms investigated.

Although for some specimens exposure to the air during the sampling procedure will be unavoidable, the biofilm must be maintained moist, because desiccation will result in collapse of the biofilm matrix. Biofilm should be fixed and/or frozen as soon as possible, preferably at the sampling site, but certainly within a few hours.

STAINING AND FIXATION (PREEMBEDDING)

Where possible, immediate immersion of the sample (substrata with attached biofilm) in a tube of chilled, fresh fixative on ice, then transportation to the laboratory for freezing is advisable to prevent sample alteration. Situations where this is not possible include the sampling of biofilm from massive substrata (e.g., pipes), which is discussed in the Sampling of Unsectionable Substrata section, and when physiological activities are to be assessed, through the uptake of radiolabeled compounds, or the use of inducible fluorescence substrates (reviewed in McFeters et al., 1995). Examples of such inducible fluorescent dyes used in environmental microbiology include fluorescein diacetate (FDA) (Swisher and Carroll, 1980); ChemchromeB and rhodamine 123 (Diaper and Edwards, 1994); sulfofluorescein diacetate (SFDA) (Tsuji et al., 1995); tetrazoliums (Altman, 1976; Blenkinsopp and Lock, 1990; Seidler, 1991; Ulrich et al., 1996; Kalmbach et al., 1997); or umbelliferyls (Hoppe, 1993). In these situations, the relevant substrate is added to the sampling tube and incubated, before the sample is fixed.

In addition, for samples that may require immunohistochemistry or enzyme histochemistry, it may be advantageous to freeze the biofilm unfixed so that the target site reactivity is not impaired. In these cases, freezing of the biofilm at the sampling site is recommended to prevent alteration of the microbial populations.

Samples must be fixed by adding sufficient concentrated, fresh paraformaldehyde (Fisons, Sydney) in phosphate-buffered saline (PBS, pH 7.2) (Oxoid, Basingstoke, Hampshire, England), to the sampling tube to achieve a final concentration of 4% paraformaldehyde, then maintain on ice or refrigerated for 1–3 hr. Paraformaldehyde is the preferred fixative when FISH probing for most Gram-negative bacteria; however, as an alternative, 50% ethanol at $-20°C$ for 3 hr is feasible and is advised for fixation when hybridizing many Gram-positive bacteria (Amann et al., 1995). Overfixation with paraformaldehyde results in poor probe binding and weaker adherence of the section to the slide. Specimens should be washed in PBS to remove excess fixative and then embedded, although some stains may be applied at this stage, prior to embedding.

Staining prior to embedding involves the addition of concentrated stain solution to a second tube of PBS wash. Staining at this stage can be advantageous when dealing with specimens that frequently become dislodged from the microscope slide during on-slide staining, after sectioning, e.g., plastic substrata. In addition to the physiological stains, common stains that may be applied at this stage include acridine orange and DAPI, as described in the Staining section. Excess stain is removed by washing in PBS for 10 min prior to embedding.

EMBEDDING (SECTIONABLE SUBSTRATA)

Many soft and semisoft substrata, such as filters, membranes, plant tissues, some plastics and some minerals, may be sectioned along with their attached biofilm. Roots and stems from wetland reeds (e.g., *Schoenoplectus validus, Phragmites australis* and *Triglochin* sp.), can be sampled for cryoembedding and cryosectioning; however, stem tissues of aquatic plants often contain aerenchyma (tissue containing air spaces) that collapse when sectioned. Because trapped air pockets within any sample will result in poor sections, steps should be taken to minimize their occurrence within the specimen. In the case of plant tissue, thin slivers (~2 mm thick) can be cut from the surface of the plant stems with razor blades or scalpels, thus avoiding the aerenchyma tissue within the stem. For samples such as hollow-fiber membrane filters, preventing the fibers from dewatering during sampling is desirable, although vacuum infusion of the embedded specimen can be beneficial where dewatering of the fiber has occurred (see below).

Care should be taken not to touch the biofilm on the slivers, which are then embedded by immersion, biofilm side down, into labeled cryomolds (intermediate size—Tissue-Tek/Miles, Elhart, IN) filled with a cryoembedding medium. Commercially prepared cryoembedding media differ between manufacturers, varying in viscosity and cryoprotectant qualities (ability to reduce ice-crystal formation). Cryoembedding media from Leica/Jung (Germany) or Tissue-Tek/Miles (OCT compound) are suitable. Other cryoprotectants such as 30% sucrose or 2–4% methylcellulose may also be used, although the commercial preparations generally give superior results. The high viscosity of the Tissue-Tek medium may be reduced by dilution with water (to ~85% of original concentration) prior to use, whereas the Leica medium can be used directly.

After immersion in embedding medium, leave specimens at room temperature for 15 minutes to allow good infiltration and then freeze, as described below. Alternatively, thick or dewatered specimens may benefit from vacuum infusion with the medium (or sucrose solution) by being immersed and then sealed within a desiccation chamber and held under vacuum for 15–30 min to ensure maximum penetration prior to freezing. The orientation of the sample should be noted on the cryomold, because the medium will be opaque after freezing, and the specimen will need to be positioned correctly when mounted for sectioning.

FREEZING

Freeze each specimen rapidly to minimize ice crystal formation, by placing filled cryomolds on a precooled metal block. Metal blocks can be precooled by placing within a −80°C freezer or by standing in a liquid nitrogen bath. Some cryostats are equipped with quick-freeze blocks (e.g., Clinicut, Bright) and these can also be used. Once frozen, samples should be sealed in plastic tubes or jars to prevent desiccation/sublimation, which would result in cracked and unsectionable blocks. Blocks can be stored at −80°C until required for sectioning (samples have been satisfactorily stored for several years).

SAMPLING OF UNSECTIONABLE SUBSTRATA

Biofilm on unsectionable substrata (e.g., pebbles, pipes and slides) can be sampled by utilizing the nature of the embedding and freezing procedure. In this case, a small area of the biofilm covered substratum (<1 sq. cm) is coated with embedding medium, which is then rapidly frozen and cracked off from the substratum, as summarized in Figure 1. To achieve this, ensure that the biofilm is moist and then either pour a small amount of embedding medium directly onto the sample or place a cryomold overfilled with embedding medium against the sample surface, in such a way that air bubbles are not trapped against the biofilm. Rapidly freeze the whole sample area as described above with precooled metal blocks or with the aid of commercial aerosol freezing sprays, such as Surefreeze (Maxwell Chemicals, Sydney), which are helpful in freezing awkward samples. Dry ice (solid carbon dioxide) (Callis et al., 1991), or liquid-nitrogen-cooled hydrocarbons (e.g., isopentane) may also prove useful; however, liquid nitrogen itself is generally not suitable because it vaporizes upon contact with the warm sample. The frozen biofilm/medium layer is then cracked off from the substratum by twisting the edge of a precooled razor blade or knife against the substratum. Clamps or pliers (also precooled) may be necessary to achieve the firm hold required for leverage. When sampling from curved or rough surfaces, keep in mind that small areas of embedding medium (<1 sq. cm) are more easily cracked off than larger areas. This cold-shock procedure cracks off the frozen biofilm, because the biofilm-substratum interface is the natural line of weakness when the specimen is frozen brittle (Costerton et al., 1986; Yu et al., 1994). The exposed base of the biofilm may be marked at this stage, with marker pen or a piece of tissue placed against the base. Invert the sample and cover the biofilm base with embedding medium so that the biofilm is sandwiched, freeze rapidly again and store as described above.

CRYOSECTIONING

The cryostat temperature should be preset so that all components (including knife, brushes, razor blades, forceps and empty slide rack) are at the desired temperature before commencing. Sectioning temperature can be critical and will vary with specimen type and brand of cryostat, and thus must be determined empirically. Temperatures between −28°C and −19°C have yielded good results for biofilm samples, with colder temperatures required for heterogeneous samples where the substratum is included (e.g., plants or filters), while warmer temperatures are suitable for biofilm specimens that have been fractured from the substratum. The Clinicut cryostat (Bright) required sectioning temperatures ~3°C lower than Leica cryostats, when using the same samples. If the temperature is too warm, compression of the section will occur in the direction of cutting; if too cold, chattering and splintering may result. Low-profile disposable blades (e.g., Feather) are suitable for some biofilms. However stronger, inflexible blades, e.g., Magna-cut disposable blades (Bright) or resharpenable steel knives, are superior for cutting heterogeneous samples of biofilm and substrata.

Before cryosectioning, place the frozen specimen into the cold cryostat chamber and allow the temperature to equilibrate (>30 min when stored at −80°C). Push the sample out of the mold and cut with a razor blade if necessary to permit the correct orientation of the biofilm (depth of the block should be <10 mm). Mount the block onto a precooled metal stub with embedding medium, section down to reveal a fresh surface of the block and then section at 4-μm thickness. Collect sections by thaw-mounting onto room temperature, silane-coated glass microscope slides (see below). Take care when thaw-mounting, to keep the slide and section parallel, thus avoiding distortion or creasing of the sections. Immediately refreeze sections on slides by retaining them in the cryostat chamber. Sections may be stored indefinitely at −80°C in a slide box containing desiccant. In some instances long coverslips may be preferred over microscope slides, because the coverslips permit microscopic visualization of the section from either side (see the In situ Microbial Activity section).

Silane-coated microscope slides may be preferred and are prepared by cleaning slides (5% HCl or 2M KOH for 30 min, distilled water rinse, 95% ethanol wash for 10 min, repeat), then air drying before dipping into fresh 2% silane (3-aminopropyl triethoxysilane) (APES) (Sigma) in acetone and rinse in acetone (Keller and Manak, 1989; Wilcox, 1993). Silane-coated slides provide greater retention of sections during histochemical procedures and are stable for several months when stored frozen with desiccant.

Cryosections should be evaluated for integrity by observation and assessment of the following:

(1) Tearing of the specimen during sectioning
(2) Compression or distortion of the section, relative to the block
(3) Integrity of structures of known morphology (plant cells, filaments, etc.)
(4) Consistency of morphology (within and between sections)
(5) Intact biofilm-substratum interface or complimentary morphology where separated

STAINING (POSTSECTIONING)

For some samples, maintaining the sections on the slide throughout staining procedures can be difficult, especially for sections containing plastic substrata (e.g., filters and membranes). Adherence can be improved through dehydration of the section (air drying and/or ethanol) or on-slide fixation (paraformaldehyde, or for some stains, heat/flame fixation).

Remove sections from the freezer and thaw-dry by placing at 50°C for 5 min to help adhere the sections to the slides (Wilcox, 1993) or first dehydrate in chilled 95% ethanol for 10 min. On-slide paraformaldehyde fixation is achieved by placing the slide in fresh paraformaldehyde fixative (as in the Staining and Fixation section) at 4°C for 30 min and then dip in tap water and dry as above, before staining or probing.

Conventional microbiological staining techniques, such as Gram and Ziehl-Neelsen (acid-fast) reactions, can be applied to the cryosections to assess cell characteristics (Doetsch, 1981), or fluorescent stains such as DAPI (4′,6-diamidino-2-phenylindole) (Sigma) and acridine orange (e.g., Sigma) provide clear visualization of even small bacterial cells. Use a fresh solution of DAPI at 1 μg ml^{-1} for 5–10 min, or acridine orange at 1 mg ml^{-1} for 2–5 min. To reduce the risk of dislodging the section from the slide, stain sections by applying and removing the stain and wash water (distilled water rather than PBS) while maintaining the slide horizontal and using a pipette.

Note: both DAPI and acridine orange bind to DNA, and are known carcinogens.

IN SITU BACTERIAL IDENTIFICATION (BY FISH)

Microscopic identification of individual bacterial cells can be achieved through the use of fluorescent in situ hybridisation (FISH), which involves on-slide incubation of the biofilm section with fluorescently labeled oligonucleotides (synthetic DNA probes) targeting ribosomal RNA (rRNA) (Amann et al., 1995). Ribosomal RNA is the molecule of choice for phylogenetic determinations because regions of both highly conserved and highly variable sequences occur and is now the basis for the reassessment of all microbiological phylogenies (Maidak et al., 1997; Olsen and Woese, 1997). This variability enables regions to be selected that can either encompass or distinguish bacterial groups at almost any phylogenetic level from domain spanning to strain specific. Such variability, combined with the occurrence of rRNA in high copy numbers per cell, makes this single stranded molecule an ideal target for FISH probing and results in an extremely powerful staining technique. Whole-cell hybridization can also be directed toward messenger RNA (mRNA), to detect the expression of particular genes. The much lower number of specific mRNA sequences per cell, however, generally necessitates the use of more sensitive detection systems such as in situ amplification techniques and/or indirect fluorescence (Hougaard et al., 1997; Komminoth and Werner, 1997; Schmidt et al., 1997).

There are many published probe sequences that target regions of the 16S and 23S rRNA, designed to discriminate between various groups of microbes, at all phylogenetic levels (Amann et al., 1995), and there are many companies that will custom synthesize, fluorescently label and purify oligonucleotide probes. For probes that have been specifically developed for whole-cell or in situ applications, the hybridization stringency required is published along with the sequence, and as such, these probes can be quickly used and tested against positive and negative control cultures. PCR primers and probes used in dot blots, however, are not always successful when applied in situ because of the tertiary structure of the intact rRNA molecule, which results in some regions being folded and inaccessible to the probes. In this situation, and for newly designed probes, empirical trial and error is required to determine if the probe can bind and the level of stringency needed. This is achieved by obtaining a signal at low formamide concentrations (i.e., low stringency) and then successively increasing the formamide level until binding of the probe is prevented. The highest formamide concentration (i.e., highest stringency) still permitting good signal strength should result in specific binding of the probe to the target sequence.

For FISH probing, silane-coated slides are used and biofilm specimens are paraformaldehyde or ethanol fixed, either before or after sectioning (see comments in the Staining and Fixation section). On-slide paraformaldehyde fixation of the specimen destroys the hydrophobic nature of silane-coated slides, causing buffer to run off the section during the hybridization. To overcome this problem when on-slide fixation is used, encircle sections by using a hydrophobic barrier such as a pap pen (Zymed, Barlingame, USA). In addition, some cell types (especially Gram-positive bacteria) may also require treatment with lysozyme.

On-slide hybridization procedures follow the approach of Manz et al. (1992), although ethanol dehydration is not necessary, nor is prehybridization or prewarming of the hybridization buffer required. For each slide to be hybridized, prepare 2 ml of hybridization buffer and 50 mL of washing buffer. The composition of both buffers depends on the probe being used. Hybridization buffer typically consists of 0.9 M NaCl, 20 mM Tris/HCl, 0.01% SDS (sodium dodecyl sulfate) and 0–50% formamide, depending on the stringency required for the probe being used (see comments above). The washing buffer contains 20 mM Tris/HCl (pH 7.4), 0.01% SDS, but no formamide, and so the NaCl concentration is modified to maintain the same stringency as the hybridization buffer, according to the formula for DNA-RNA hybrids (Wahl et al., 1987). This relationship halves the salt concentration in the washing buffer, with each 10% increase in the formamide concentration used in the hybridization buffer, progressing from 0.90 M NaCl for 0% formamide, to 0.03 M NaCl for 50% formamide.

Always handle probes, solutions and samples while wearing gloves to reduce contamination of the sample with RNase from the skin. Apply 1 µl of each fluorescently labeled oligonucleotide probe (30–50 ng ml^{-1}) and 8 µl of hybridization buffer onto a small area of the biofilm section (<1 sq. cm). The probe and buffer may either be mixed in a tube prior to application or directly on the slide surface, by using a pipette tip and taking great care not to touch the underlying section. Place the slide gently into a 50-ml centrifuge tube containing a thin wad of absorbent paper moistened with 2 ml of hybridization buffer (this will humidify the tube). Hybridize by sealing the tube and incubating at 46°C for 1.5 hours. After hybridizing, quickly transfer slides into 50 mL of prewarmed washing buffer and leave at 48°C for 15–20 min. If the sample temperature drops significantly during this transfer, non-specific binding of the probe may result. After washing, gently rinse slides with distilled water to remove salts and then counter stain with DAPI (see in the Staining section) and allow to air dry in darkness before mounting and viewing. When using fluorescein-labeled probes, specimens should be mounted in an antifading solution, such as Vectorshield (Vector Laboratories, San Francisco, CA). For other fluorophores, mount specimens in low-fluorescence immersion oil (e.g., type FF, Cargille, Cedar Grove, NJ). Appropriate positive and negative control cultures should be used on the slides to confirm the specificity of probe binding.

Probes requiring different formamide concentrations can be used on the same sample; however, this requires separate, successive hybridizations. The probe needing the higher formamide concentration should be applied, hybridized and washed, before applying the second probe for a second, less stringent hybridization and wash.

Note: formamide and most antifading solutions are known carcinogens.

IN SITU MICROBIAL ACTIVITY (BY MICROAUTORADIOGRAPHY)

The location of radioactively labeled compounds within biofilm cryosections can be visualized through the use of microautoradiography. This approach involves covering the biofilm section with a thin layer of sensitive emulsion in which silver crystals form when energized by the radioactivity (or light). Because the silver grains are superimposed over the specimen, the two may be viewed simultaneously, thus permitting radioactive regions to be identified and correlated with biological structures, e.g., cellular uptake of radiolabeled compounds that were incubated with the living biofilm. When silver grains in the emulsion are densely clustered, they can obscure observation of the fluorescently stained specimen below; however, if the sections were initially thaw-mounted onto silane-coated coverslips rather than microscope slides, the emulsion-covered sample can be sandwiched between two coverslips, permitting visualization from either side. Compounds commonly labeled isotopically for physiological studies include carbon sources (e.g., acetate, carbon dioxide), nucleotide precursors (e.g., thymidine) and protein precursors (e.g., leucine) (Robarts and Zohary, 1993; Andreasen and Nielsen, 1997).

Thaw-dry at 50°C for 5 min cryosections of radiolabeled biofilm on silane-coated slides or silane-coated coverslips and then coat with NTB-2 emulsion (Kodak) following the manufacturer's instructions (Eastman Kodak, 1995) and the methodology of Carman (1993), which is summarized below. Until development is complete, it is preferable to keep the emulsion protected from all light sources (including static-electric sparks), although darkroom safe-lights have been used briefly by some workers, after the slides have been coated (Eastman Kodak, 1995). Heat the emulsion to 43°C and dilute 2 parts emulsion to 1 part distilled water. Dip the slides into the melted emulsion, and allow to gel and air dry horizontally, protect from light, seal in an airtight container and leave to expose for days to weeks, depending on the signal strength. A variety of control slides with and without exposure to isotope should also be prepared to assess for the occurrence of positive or negative chemography (chemical energizing of the emulsion). Exposure time is ascertained by periodically processing sacrificial slides to check the development of silver grains in the emulsion. When sufficient response is observed, develop and fix the slides with Dektol developer and standard fixer (Kodak), according to the manufacturer's instructions and then either counterstain immediately or allow to air dry and refrigerate with desiccant until required.

The microautoradiography sample can also be stained, either before or after coating wih emulsion. Staining of the sample before covering with emulsion is preferable, because this gives clearer staining with lower backgrounds; however, some stains can energize the emulsion, while others (e.g., fluorescein-labeled antibodies or oligonucleotides) may fade during lengthy exposure times and can be affected by the alkaline nature of the emulsion. Such chemical interference between emulsion and stain may be reduced by placing a barrier layer (such as silane) over the stained specimen, before dipping into the emulsion. Alternatively, after the exposure and development, the specimen can be stained through the emulsion, as described for DAPI or acridine orange in the Staining section, provided that samples are rinsed thoroughly.

IN SITU CHEMICAL MAPPING (BY EDX)

Chemical components of biofilms can be located and characterized by the EDX (energy dispersive X-ray analysis) facility provided with many scanning electron microscopes (SEMs). The electron beam of an SEM is used to excite X-ray emissions from elements within the sample. Each element produces X-ray photons with characteristic energies, from which it can be identified. EDX does not detect low atomic weight elements, including carbon, nitrogen, and oxygen, because they do not interact sufficiently with the electron beam to produce a strong signal (i.e., they are "electron transparent"), and the low energy X-rays produced (<1 keV) are masked by instrumental artifacts (Russ, 1984). Furthermore, EDX gives no information about chemical bonding (valence states) of the elements detected. The specimen can be simultaneously imaged by the SEM, allowing selective analysis of small areas of the specimen by controlling the area of the beam raster. The depth to which the specimen is analyzed depends on its electron transparency but is generally limited to the uppermost few micrometers.

For EDX analysis, the biofilm cryosection must be thaw-mounted onto an organic support. Glass slides will give high background peaks (chiefly silicon). Suitable supports include nitrocellulose membrane filters (such as Millipore HA). The filter and specimen are mounted with colloidal carbon glue onto a graphite stub (Ted Pella Inc., Redding, CA). After drying the stub is coated by using a carbon arc (Balzers), producing a conductive blanket over the specimen to drain electrostatic charge induced by the electron beam during imaging. Metal films are unsuitable as they will appear in the analysis. The specimen is then examined by SEM. X-rays generated by beam interactions are sorted by an energy-dispersive detector and pulse processor (e.g., Kevex model 4461A). Characteristic emission peaks in the energy range 0–10 keV will identify the range of elements present. By controlling the imaging area, composition can be determined at a single point or averaged for an area of interest. Advanced software packages (e.g., Iridium, IXRF Systems Inc., Houston, TX) facilitate the production of raster intensity maps, or linear transects, showing the distribution of elements of interest within the specimen. Elemental content can be quantified (as % weight of the total of detectable elements only) by comparison with standards; some software packages also offer standardless analysis.

RESULTS

MORPHOLOGY AND BASIC STAINING

Good quality, intact sections have been obtained from all of the sample types listed as suitable in the section, Range of Applications. Only samples of granular activated carbon were unsuitable, because the sectioned granules dispersed when thaw-mounted onto microscope slides, frequently disrupting the adherent biofilm. In agreement with the study of artificial biofilm (Yu et al., 1994), fracturing of frozen samples results in clean separation along the biofilm-substratum interface, as can be seen in Figures 2(a) and 2(b), permitting the biofilm structure to be preserved while removing the sample from the substrata. Environmental biofilms containing particulates were successfully sectioned with no apparent disruption of biofilm structure in the vicinity of particles.

Sections of biofilm cut at ≤ 5 μm thick clearly reveal biofilm structures and channel morphology (see Figure 2(b)) and are not significantly altered by sub-

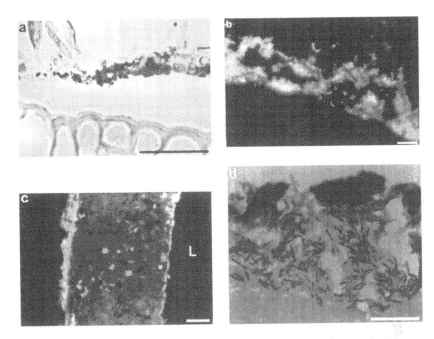

Figure 2 Photomicrographs of cryosectioned environmental biofilms—simple stains. Scale bars = 20 μm. Sections are 4-μm-thick cross sections. Voids (marked with arrows) were confirmed under phase contrast microscopy. (a) Unstained epiphytic, wetland biofilm can be seen fracturing cleanly away from the substratum, whereas the embedded biofilm itself maintains structural integrity. (b) and (c) Polypropylene hollow fibers filtering screened primary sewage and frequently backwashed. (b) DAPI stained, highly structured bacterial biofilm from the exterior of a hollow fiber microfilter. Note the curved biofilm base resulting from the cylindrical substratum (removed). (c) Acridine orange stained hollow-fiber microfilter cross section, showing biofilm growth on both pre- and postfiltration surfaces and bacterial colonies within the microfilter itself. The lumen is labeled L; flow is from the outside to the lumen. (d) Intact biofouling layer from a reverse osmosis membrane filter, treating filtered, chlorinated, sewage. Acid-fast staining mycobacteria can be seen within the biofilm.

sequent specimen dehydration, as required in many staining procedures. Thicker sections (10–30 μm), however, are best viewed by confocal microscopy, while still hydrated, to prevent the biofilm collapsing against the slide and obscuring morphological detail.

The relative amounts and location of particulates, algae and bacteria can be observed in unstained biofilm cryosections (see Figure 2(a)) although superior images of bacterial cells are achieved when DNA-binding stains such as DAPI or acridine orange are used (Figures 2(b) and 2(c), respectively).

Biofilm morphology was observed to vary widely between samples, ranging from densely packed cells with few pore spaces, to highly porous films with cell clusters of varying densities (see Figures 2 and 3). In some biofilm

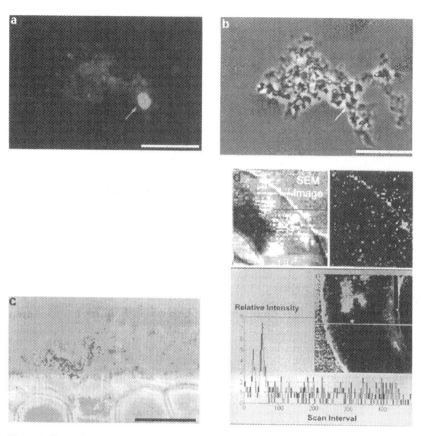

Figure 3 Photomicrographs of cryosectioned environmental biofilms—complex stains. Scale bars = 20 μm. Sections are 4-μm-thick cross-sections. (a) and (b) Biofilm from a chloraminated drinking water system, imaged after fluorescent in situ hybridization (FISH), with oligonucleotides targeting ribosomal RNA (rRNA), to identify ammonia-oxidizing bacteria (arrow), which can be seen occurring in a tight cluster. Same field shown under phase contrast (a) and epifluorescence microscopy (b). (c) Microautoradiography of epiphytic wetland biofilm, incubated in situ with [methyl-³H]thymidine. Silver grains in the overlying emulsion indicate clustered localization of the isotope within the unstained biofilm below. This image is focused for the emulsion, leaving the specimen slightly out of focus. (d) Polypropylene hollow fiber filtering screened primary sewage and frequently backwashed. A quadrant of cross-sectioned fiber is chemically assessed by energy dispersive X-ray analysis (EDX) and analytical software (Iridium), to produce a raster map, indicating the distribution of sulfur (top right). An SEM secondary electron emission image was made simultaneously (top left). Another section is similarly analyzed to produce a quantitative, graphical display of sulfur concentration along a linear transect, rather than a raster map.

specimens the thickness varied sharply within each sample. This was most extreme in drinking water biofilms where a sparse monolayer could suddenly give rise to columns of 300 μm in height. Preparation of hollow fiber membranes by cryosectioning was particularly informative because it provided images of biofouling on both sides of the nominal 0.2 μm filter, together with structural damage and bacterial growth within the filter, as can be seen in Figure 2(c).

Morphological parameters, such as cell density, cluster size, porosity, surface roughness and nearest neighbour calculations, can be more easily assessed by using digitized images and image analysis software.

In Figure 2(d), acid-fast staining has been used to locate microcolonies of mycobacteria within a biofilm fouling a reverse osmosis membrane used in wastewater treatment. This demonstrates the ability of traditional staining techniques to provide spatial information on microbial populations when combined with cryosectioning.

IN SITU BACTERIAL IDENTIFICATION (BY FISH)

Comparison of Figure 3(a) and 3(b) demonstrates the ability of FISH probing to easily identify individual cells that are otherwise impossible to recognize. In this example, individual bacterial cells are identified as ammonia-oxidizing bacteria by FISH with a CY3-labeled probe (NEU23a) (Wagner et al., 1995) binding to a group of tightly clustered cells within biofilm from a chloraminated drinking water system.

Hybridization with several probes, each labeled with a different fluorophore, allows powerful multicolored probing to simultaneously identify cells of several different phylogenetic groups, which can then be enumerated and their spatial relationships determined.

IN SITU MICROBIAL ACTIVITY (BY MICROAUTORADIOGRAPHY)

Microautoradiography provides an overlay above the biofilm cross section, where silver grains indicate regions of isotope incorporation. Visualization of the biofilm structure and staining of cells in the underlying biofilm is possible. Figure 3(c) shows an example of microautoradiography as applied to cryosectioned epiphytic biofilm, incubated with tritiated thymidine prior to embedding. Silver grains in the emulsion show the location of cells that incorporated the radiolabeled thymidine into new DNA during the incubation period, indicating that cellular growth was occurring within this cluster.

IN SITU CHEMICAL MAPPING (BY EDX)

Figure 3(d) shows an EDX analysis of a fouled hollow fiber microfilter membrane used for water reclamation from sewage effluent. A quadrant of a transverse section of the fiber wall is shown imaged by SEM and mapped for sulfur, aluminium and phosphorus. Sulfur (in this case shown by GC-MS to be elemental sulfur (S_8)) is strongly concentrated in the surface fouling layer. The use of cryosectioning has overcome a traditional weakness of EDX analysis, i.e., its inability to return information from below the outermost surface layer.

COMPARISON WITH OTHER METHODS

Of the environmental studies that have addressed biofilms, most have relied on traditional microbiological techniques such as heterotrophic plate counts as applied to dispersed biofilm scrapings (e.g., Morikawa, 1988; Chand et al., 1992). Dispersion and culturing techniques ignore biofilm morphology and introduce a huge culture-dependent bias into population analyses, thus rendering them largely unsuitable for environmental applications (Wagner et al., 1993). Recent investigations have utilized an array of techniques, searching for methods that can accommodate biofilms and produce meaningful results. Among the techniques investigated are whole-community exoenzyme analysis (Sinsabaugh et al., 1991), phospholipid analysis (Scholz and Boon, 1993) and genetic profiles (Moyer et al., 1994). Few of these studies, however, have included biofilm morphology or provided information on how biofilms function.

Structure is critical in understanding the functioning of biofilm communities, because it controls the distribution of microenvironments, nutrient availability and hence physiological activity (Costerton, 1994; Costerton et al., 1994). Scanning electron microscopy has been frequently used in biofilm structural investigations, yet it results in images only of the surface topography of the sample, revealing nothing about the inner structure and organization of the biofilm. This drawback, combined with the artifacts introduced by harsh dehydration and sample preparation procedures, have cast serious doubt over the ability of SEM of whole biofilm to provide useful and realistic information about in situ structure. Frozen and environmental SEM are two techniques that circumvent the necessity for dehydration and extensive sample preparation. Both of these approaches have been applied to biofilms and may provide high-resolution images of hydrated, undisrupted samples (Sutton et al., 1994; Hodgson et al., 1995). However, as with traditional SEM, these methods only provide information about the very surface of the biofilm and are unlikely to become daily tools in biofilm monitoring because of their expense and limited availability.

Confocal laser scanning microscopy (CLSM) of laboratory grown biofilms has revealed heterogeneous structures of cell clusters interspersed with voids and channels (Lawrence et al., 1991; deBeer et al., 1993; Stoodley et al., 1994). These structures have been shown to alter with species composition and environmental conditions such as nutrients, flow rates and the presence of inhibitors (Lawrence et al., 1991; Huang et al., 1995; Møller et al., 1997; Stoodley et al., 1997). Thus, confocal microscopy can provide excellent, detailed information about biofilm structure and function, which will no doubt continue to increase our understanding of biofilms. Confocal microscopy is, however, expensive in terms of equipment and labor required (time and skill). In addition, confocal microscopy is difficult to apply to most environmental biofilms, because of the uneven and often massive nature of their substrata and the incorporation of particles. The very act of sampling the biofilm usually destroys its integrity and even if intact, thick environmental films often contain too much visual information to allow fast, easy viewing and interpretation, even with CLSM (Wagner et al., 1994; Wagner et al., 1996). These problems are also encountered in many medical samples, where they have been largely overcome through the use of thin sectioning.

Thin sectioning has been applied to biofilms, particularly medical biofilms from catheter infections, the gut and teeth (Costerton et al., 1986; Costerton, 1994). In addition, some environmental biofilms have also been thin sectioned; however, these investigations have utilized traditional preparation and embedding procedures, especially those associated with transmission electron microscopy (TEM) (Lappin-Scott and Costerton, 1989; Väisänen et al., 1994). These standard histological techniques, such as resin and paraffin embedding, are labor intensive, requiring long preparation times (days to weeks) and dehydration of the sample and the use of various, often toxic chemicals for fixation and contrast enhancement. These procedures commonly result in the reduced reactivity of immunological, enzymatic and genetic target sites that contribute to poor specific staining (Wilcox, 1993). In addition, paraffin sections require dewaxing and dehydration, which restricts the combined use of pre- and post-embedding stains, because the preparation procedures can easily remove stains applied prior to embedding (e.g., physiological stains).

In medical situations such difficulties are frequently circumvented via the use of cryoembedding and subsequent cryosectioning. Freezing and cryosectioning have significant benefits over other forms of sectioning. Because cryoembedding rapidly freezes the specimen, fixation is not required to prevent alterations within sample, whereas fixation is essential prior to other embedding procedures because of their long preparation times. As such, unfixed cryosections retain the reactivity of histological target sites to a much greater degree than other section types. Cryoembedding requires no dehydration of the specimen, allows rapid preparation using an inert aqueous medium, and is compatible with standard histological, immunological, enzymatic and genetic techniques. Furthermore, frozen samples may be stored indefinitely (Wilcox, 1993), and each sample can produce hundreds of sections, permitting a vast array of stains and probes to be applied to the same sample and viewed within the intact morphology of the specimen.

The aqueous nature of cryoembedding media, along with the fast preparation time, ensures that the medium supports the external structures of the specimen but does not permeate throughout the cells. This results in cryoembedded specimens retaining their original texture and structural heterogeneity, unlike the more invasive resin embedding media used in TEM, which penetrate and bind together all elements of the sample. Consequently, some specimens possessing heterogeneous density are not well suited to cryosectioning. Cryosectioning has been successfully applied to artificial biofilms (Yu et al., 1994); however, the technique was considered unsuitable for environmental samples and was rejected as being of limited use with such specimens because of their thickness and the presence of abiotic, rigid and fibrous material (Huang et al., 1996). Hence, these preconceptions prevented the application of cryosectioning to environmental biofilms.

In agreement with Huang et al. (1996), we also found the methods previously published for cryosectioning biofilms to be unsuccessful in environmental samples. Thus, cryoembedding and cryosectioning techniques were successfully optimized for application with a wide range of biofilms and fouling layers. Optimization procedures involve the use of colder temperatures, stronger blades, sample infusion and selective tissue sampling. These approaches were successful with samples derived from a variety of environments and substrata, spanning aerobic and anaerobic conditions, abattoir and sewage wastewaters, river and drinking water systems, exhumed pipes, reverse osmosis membranes, hollow-fiber microfilters, river rocks, plants and plastic supports.

DISCUSSION

There is currently little information available about the spatial relationships of bacterial species in mature environmental biofilms, and hence many questions to answer, the main obstacle to which has been a lack of suitable techniques with which to deal with biofilm structure. Most of these questions could be explored in natural, environmental biofilms with the application of cryosectioning. Once biofilm has been cryosectioned, it may be analyzed in a variety of ways; indeed, the same sample yields dozens of sections, each of which may be analyzed differently. Associations and spatial relationships can be investigated by recording the relative positions and frequencies of various specific cell types, either phylogenetically or physiologically. The potential of the FISH technique is considerable and will surely provide new insights in the area of spatial relationships. The impact of biofilm structure can be considered and structural quantification may be determined manually or with the aid of image analysis of a digitized image (Lawrence et al., 1991).

Cryosectioning provides not only the opportunity to observe biofilm structure quickly and cheaply with multiple histochemical staining per sample, but the cryoembedding approach also represents a unique method of sampling biofilms from the environment. Preparation of environmental biofilms by cryosectioning facilitates the collection of in situ biofilm data and can thereby increase our knowledge of these prevalent microbial structures.

SENSITIVITY AND ACCURACY

FISH

Direct assessment of cell types by FISH avoids any averaging of the sample, thus providing very precise information, at high resolution, of the nature of the individual populations and their possible role in the biofilm as a whole. Image analysis software can be applied to biofilm sections, enabling manipulation of background and autofluorescence levels, relative to probe signal intensity, thus enhancing specific structures of interest (Wagner et al., 1994). FISH probing of rRNA can be used to identify individual cells at any phylogenetic scale, from the broad domain level, down to the subspecies strain.

Positive and negative control cultures and control hybridizations (lacking probe) should always be checked to confirm specificity of probe binding and development of autofluorescence during the procedures. Where only a single base difference exists between the target sequence and a non-target sequence, it is advisable to co-hybridize with an unlabeled competitor probe, designed to bind to the non-target sequence. For example, probes Bet42a and Gam42a (targeting the Beta and Gamma *Proteobacteria*) (Manz et al., 1992), should be hybridized in conjunction with, and under stringent conditions, to prevent cross binding.

In general, hybridization-conferred fluorescence is scored on a positive or negative basis for each cell; however, some investigators have used fluorescence intensity measurements to estimate ribosome concentrations within a cell and extrapolate this as an indicator of cellular activity (Poulsen et al., 1993). The degree of correlation between activity and rRNA content, however, may be species specific (Flärdh et al., 1992; Hsu et al., 1994; Fukui et al., 1996).

Microautoradiography

Microautoradiography is suitable for many isotopes, and there are different types of emulsion, each tailored for various isotopes and sensitivities. When the appropriate emulsion is selected, extremely high sensitivity can be achieved by extended exposure times. When used with microorganisms, microautoradiography is generally used to identify individual cells as positive or negative for the possession of isotopes, and because there is always some degree of background (energized grains), a threshold number of silver grains is set, which must be attained by each cell to be enumerated as positive. The resolution of the emulsion, however, depends on the inherent grain size of the emulsion (usually >0.3 μm) and cannot be altered. When used with densely clustered biofilm cells, the grain size of the emulsion, combined with the need for a threshold number of silver grains, may give insufficient resolution to identify individual cells as either positive or negative. However, whole clusters or regions in the biofilm can be assessed. The grain size also limits the minimum thickness of emulsion that may be used without reduced sensitivity.

EDX

EDX allows qualitative and quantitative assessment of the elemental content of a specimen, with spatial resolution appropriate for bacterial habitats such as biofilms. EDX is best applied to determine the nature of "abiotic" inclusions, such as mineral grains. As mentioned in the section In situ Chemical Mapping, wholly organic biological materials will not return elemental information, with the possible exception of phosphorus. Direct structural interpretation is also problematic. Recognizable biological structures, such as cells, may collapse during dewatering, and, because the specimen is not metal coated, the SEM image is of low quality. Metal-coated specimens can be analyzed if the elemental peaks of interest do not overlap with those of the coating metal (gold peaks, for example, obscure sulfur). It is probably best to locate biological structures by light microscopy of an adjacent section.

HANDLING AND MAINTENANCE

Because biofilm populations can rapidly respond to environmental changes (Wagner et al., 1993; Sutton et al., 1994), immediate freezing and/or fixation is required to prevent any alteration in the population structure. It should be remembered that freezing is not a fixation/disinfection step, and thus, unfixed samples should be handled carefully because they may contain viable pathogenic organisms.

LONG-TERM STABILITY AND FIELD STABILITY

Once fixed and frozen, biofilm samples can be stored indefinitely and remain suitable for staining, FISH probing or EDX. The half-life of any isotope used will determine the length of storage possible for those samples.

LIMITATIONS

Cross sectioning provides a very detailed view of a relatively small area of biofilm and, therefore, needs sound replication to ensure representative sampling and observation.

Some samples, such as soil and activated carbon granules, are not suitable for cryosectioning, and some samples will have high background autofluorescence that may interfere with FISH analysis, although this can sometimes be reduced by heat and chemical treatments, or circumvented through the use of infrared emitting fluorophores and digital image analysis techniques.

Software for the microscopic enumeration of cells has been largely designed for cell suspensions and may not accurately enumerate cells within tight biofilm clusters (Busscher et al., 1991; Bloem et al., 1992).

Microautoradiography can be a slow process, depending on the isotope and signal strength, and can take days to weeks of exposure to achieve the latent image.

EDX cannot detect low molecular weight elements nor determine valence states.

COSTS

Cryosampling and sectioning is inexpensive for consumables and labor time, and access to a cryostat and technician is readily available within research institutions and hospitals.

FISH consumable costs are moderate, if many samples are to be assessed by using the same batch of probe, because the minimum amount supplied is sufficient for hundreds of hybridizations. Fluorescence microscopes are common, although time commitments are high and some technical skill is required for both the hybridizations and microscopy.

Microautoradiography is expensive for consumables, because of both the isotope and emulsion. Access to a photographic dark room and training in the handling of isotopes are required.

EDX analysis costs are moderate. Suitable microscopists and/or technicians are generally available at research institutions on hourly charge-out rates, for sample preparation and imaging.

NOMENCLATURE

CLSM:	Confocal laser scanning microscopy
DAPI:	4',6-Diamidino-2-phenylindole
DNA:	Deoxyribonucleic acid
EDX:	Energy dispersive X-ray analysis
FDA:	Fluorescein diacetate
FISH:	Fluorescent in situ hybridization
GAC:	Granular activated carbon
GC-MS:	Gas chromatography-mass spectrometry
keV:	Kiloelectron volts (an energy unit, $1\ eV = 1.06^{-19}$ joules)
mRNA:	Messenger ribonucleic acid
PCR:	Polymerase chain reaction
rRNA:	Ribosomal ribonucleic acid

SDS: Sodium dodecyl sulfate
SEM: Scanning electron microscopy
TEM: Transmission electron microscopy

ACKNOWLEDGEMENTS

This work was funded by the CRC for Waste Management and Pollution Control Limited, a center established and supported under the Australian Government's Co-operative Research Centres Program.

Gayle Callis, at Montana State University, kindly demonstrated cryosectioning of artificial biofilms and provided helpful suggestions. Yung Dai, in the School of Microbiology and Immunology, University of New South Wales (UNSW), generously gave her time and advice when teaching cryosectioning. Michael Wagner, from the Technical University of Munich, generously gave advice on FISH probing, and prepublication access to FISH probes targeting nitrifying bacteria. Elizabeth Eager and AWT-EnSight kindly provided drinking water biofilms. Students at UNSW provided an assortment of biofilm support matrices from their reactor systems.

REFERENCES

Alman, F.P. 1976. *Prog. Histochem. Cytochem.* 9, 1.

Amann, R.I., W. Ludwig, K.-H. Schleifer. 1995. *Microbiol. Rev.* 59, 143.

Andreasen, K. and P.H. Nielsen. 1997. *Appl. Environ. Microbiol.* 63, 3662.

Blenkinsopp, S.A. and M.A. Lock. 1990. *Water Res.* 24, 441.

Bloem, J., D.K. van Mullem and P.R. Bolhuis. 1992. *J. Microbiol. Methods* 16, 203.

Busscher, H.J., J. Noordmans, J. Meinders and H.C. van der Mei. 1991. *Biofouling* 4, 71.

Callis, G.M., M. Jutila and S. Kurk. 1991. *Histo-Logic* 21, 253.

Carman, K. 1993. Microautoradiographic detection of microbial activity. In: P.F. Kemp, B.F. Sherr, E.B. Sherr and J.J. Cole (eds.): *Handbook of methods in aquatic microbial ecology.* Lewis Publishers, London, 397.

Chand, T., R.F. Harris and J.H. Andrews. 1992. *Appl. Environ. Microbiol.* 58, 3374.

Costerton, J.W. 1994. Structure of biofilms. In: G.G. Geesey, Z. Lewandowski and H.C. Flemming (eds.): *Biofouling and biocorrosion in industrial water systems.* Lewis Publishers/CRC Press, Boca Raton, FL, 1.

Costerton, J.W., J.C. Nickel and T.I. Ladd. 1986. Suitable methods for the comparative study of free-living and surface associated bacterial populations. In: J.S. Poindexter and E.R. Leadbetter (eds.): *Bacteria in nature, Vol. 2,* 49.

Costerton, J.W., Z. Lewandowski, D. deBeer, D. Caldwell, D. Korber and G. James. 1994. *J. Bacteriol.* 176, 2137.

de Beer, D., P. Stoodley, F. Roe and Z. Lewandowski. 1993. *Biotech. Bioeng.* 43, 1131.

Diaper, J.D. and C. Edwards. 1994. *J. Appl. Bacteriol.* 77, 221.

Doetsch, R.N. 1981. Determinative methods of light microscopy. In: P. Gerhardt, R.G.E. Murray, R.N. Costinlow, E.W. Nester, W.A. Wood, N.R. Krieg and G.B. Phillips (eds.): *Manual of methods for general bacteriology.* American Society for Microbiology, Washington, DC, 21.

Eastman Kodak Co. (ed.). 1995. *Microautoradiography: autoradiography at the light microscope level.* Kodak, New Haven, CT.

Flärdh K., P.S. Cohen and S. Kjelleberg. 1992. *J. Bacteriol.* 174, 6780.

Fukui, M, Y. Suwa and Y. Urushigawa. 1996. *FEMS Microbiol. Ecol.* 19, 17.

Hodgson, AE, S.M. Nelson, M.R.W. Brown and P. Gilbert. 1995. *J. Appl. Bacteriol.* 79, 87.

Hoppe, H.-G. 1993. Use of fluorogenic model substrates for extracellular enzyme activity (EEA) measurement of bacteria. In: P.F. Kemp, B.F. Sherr, E.B. Sherr, and J.J. Cole (eds.): *Handbook of methods in aquatic microbial ecology.* Lewis Publishers, London, 423.

Hougaard, D.M., H. Hansen and L.I. Larsson. 1997. *Histochemistry* 108, 335.

Hsu, D., L.-M. Shih and Y.C. Zee. 1994. *J. Bacteriol.* 176, 4761.

Huang, C.-T., F.P. Yu, G.A. McFeters and P.S. Stewart. 1995. *Appl. Environ. Microbiol.* 61, 2252.

Huang, C.-T., F.P. Yu, G.A. McFeters and P.S. Stewart. 1996. *Biofouling* 9, 269.

Kalmbach, S., W. Manz and U. Szewzyk. 1997. *FEMS Microbiol. Ecol.* 22, 265.

Keller, G.H. and M.M. Manak. 1989. Preparation of cells and tissues for in situ hybridization. In: G.H. Keller and M.M. Manak (eds.): *DNA probes.* Stockton Press, New York,. 59.

Komminoth, P. and M. Werner. 1997. *Histochemistry* 108, 325.

Lappin-Scott, H.M. and J.W. Costerton.1989. *Biofouling* 1, 323.

Lawrence, J.R., D.R. Korber, B.D. Hoyle, J.W. Costerton and D.E. Caldwell. 1991. *J. Bacteriol.* 173, 6558.

Maidak, B.L., G.J. Olsen, N. Larsen, R. Overbeek, M.J. McCaughey and C.R. Woese. 1997. *Nucleic Acids Res.* 25, 109.

Manz, W., R. Amann, W. Ludwig, M. Wagner and K.-H. Schleifer. 1992. *Syst. Appl. Microbiol.* 15, 593.

McFeters, G.A., F.P. Yu, B.H. Pyle and P.S. Stewart. 1995. *J. Microbiol. Methods* 21, 1.

Møller, S., D.R. Korber, G.M. Wolfaardt, S. Molin and D.E. Caldwell. 1997. *Appl. Environ. Microbiol.* 63, 2432.

Morikawa, K. 1988. *Microbial Ecol.* 15, 217.

Moyer, C.L., F.C. Dobbs and D.M. Karl. 1994. *Appl. Environ. Microbiol.* 60, 871.

Olsen, G.J. and C.R. Woese. 1997. *Cell* 89, 991.

Poulsen, L.K., G. Ballard and D.A. Stahl.. 1993. *Appl. Environ. Microbiol.* 59, 1354.

Robarts, R.D. and T. Zohary. 1993. *Adv. Microbial Ecol.* 13, 371.

Russ, J.C. 1984. *Fundamentals of energy dispersive X-ray analysis.* Butterworths, London.

Schmidt, B.F., J. Chao, Z.G. Zhu, R.L. Debiasio and G. Fisher. 1997. *J. Histochem. Cytochem.* 45, 365.

Scholz, O. and P.I. Boon. 1993. *Hydrobiologia* 259, 169.

Seidler, E. 1991. *Prog. Histochem. Cytochem.* 24, 1.

Sinsabaugh, R.L., D. Repert, T. Weiland, S.W. Golladay and A.E. Linkins. 1991. *Hydrobiologia* 222, 29.

Stoodley, P., D. de Beer and Z. Lewandowski. 1994. *Appl. Environ. Microbiol.* 60, 2711.

Stoodley, P., A. Cunningham, Z. Lewandowski, J. Boyle and H. Lappin-Scott. 1997. *Proceedings of the International Conference on Biofilms in Aquatic Systems,* Royal Society for Chemistry. Coventry, UK.

Sutton, N.A., N. Hughes and P.S. Handley. 1994. *J. Appl. Bacteriol.* 76, 448.

Swisher, R. and G.C. Carroll. 1980. *Microbial Ecol.* 6, 217.

Tsuji, T., Y. Kawasaki, S. Takeshima, T. Sekiya and S. Tanaka. 1995. *Appl. Environ. Microbiol.* 61, 3415.

Ullrich, S., B. Karrasch, H.-G. Hoppe, K. Jeskulke and M. Mehrens. 1996. *Appl. Environ. Microbiol.* 62, 4587–4593.

Väisänen, O.M., E.-L. Nurmiaho-Lassila, S.A. Marmo and M.A. Salkinoja-Salonen. 1994. *Appl. Environ. Microbiol.* 60, 641.

Wagner, M., R. Amann, H. Lemmer and K.-H. Schleifer. 1993. *Appl. Environ. Microbiol.* 59, 1520.

Wagner, M., B. Assmus, A. Hartmann, P. Hutzler and R. Amann. 1994. *J. Microscopy* 176, 181.

Wagner, M,. G. Rath, R. Amann, H.-P. Koops and K.-H. Schleifer. 1995. *System Appl. Microbiol.* 18, 251.

Wagner, M., G. Rath, H.-P. Koops, J.A. Flood and R. Amann. 1996. *Water Sci. Technol.* 34, 237.

Wahl, G.M., S.L. Berger and A.R. Kimmel. 1987. *Methods Enzymol* 152, 399.

Wilcox, J.N. 1993. *J. Histo. Cytochem.* 41, 1725.

Yu, F.P., G.M. Callis, P.S. Stewart, T. Griebe and G.A. McFeters. 1994. *Biofouling* 8, 85.

CHAPTER 7

Development of a New Method to Determine the Metabolic Potential of Bacteria in Drinking Water Biofilms: Probe Active Counts (PAC)

SIBYLLE KALMBACH
WERNER MANZ
ULRICH SZEWZYK

INTRODUCTION

D RINKING water and the adherent microbial populations forming biofilms in the distribution system is one of the most extensively studied oligotrophic habitats (Ridgway and Olson, 1981; Costerton et al., 1986; LeChevallier et al., 1987; Pedersen, 1990). The general structure of drinking water biofilms has been studied by using light or scanning electron microscopy (McCoy et al., 1981; Costerton et al., 1986), and several biofilm-forming bacteria have been characterized and classified by using conventional culture techniques (Costerton et al., 1986; LeChevallier et al., 1987). In oligotrophic systems, however, only 0.01–1% of the total bacterial community can be cultivated by using these techniques, which has been described as "the great plate count anomaly" by Staley and Konopka (1985). Most of these bacteria could be either dead, or cells are principally viable and active in their natural environment (Staley and Konopka, 1985; Roszak and Colwell, 1987) but fail to form colonies under standard laboratory conditions. Applying a recent definition of bacterial death, arguing that death of individual cells leads to the loss of morphological integrity and that dead cells cannot be stained with DAPI or acridine orange (Gonzales et al., 1992), the presence of large numbers of dead bacteria in a natural ecosystem seems to be highly unlikely. However, the inability of most cells to form colonies might be due to bacteria having entered various physiological states, such as the viable but nonculturable (VBNC) state (Morita, 1985), in response to changing environmental conditions. In addition, some bacterial species might not grow because adequate culture conditions have not been developed yet.

Fluorescent in situ hybridization (FISH) using 16S- and 23S-rRNA targeted oligonucleotide probes provides a useful tool for the identification and enumeration of bacteria within their natural habitats (Amann et al., 1995) and the simultaneous assessment of their metabolic potential. The hybridization procedure is based on the reversible binding of short, synthetic, fluorescently labeled oligonucleotides to their complementary target sites on the intracellular rRNA, which serves as a molecular marker molecule and displays variable and hypervariable and highly conservative sequence stretches. This different degree of evolutionarily constancy within the rRNA molecule enables the development of diagnostic in situ probes on different phylogenetic levels ranging from domains to species (Amann et al., 1990; Manz et al., 1992; Manz et al., 1996). In addition to the inherent phylogenetic information of the rRNA, oligonucleotide probes can be used to assess the metabolic potential of single bacterial cells, based on the correlation between growth rate, cellular ribosome content and the resulting intensity of the hybridization signal (Schaechter et al., 1958; DeLong et al., 1989; Poulsen et al., 1993). Because the identification of bacterial cells by in situ probing does not require the prior cultivation of the investigated organisms, this method can be used to determine the amount of all metabolically active bacteria in a certain microenvironment including the principally culturable species possibly present in a VBNC-state.

The reduction of the fluorochrome 5-cyano-2,3-ditolyl tetrazolium chloride (CTC) as a parameter of bacterial respiratory activity has been introduced by Rodriguez et al. (1992) and has been widely used in different ecosystems (Schaule et al., 1993; Coallier et al., 1994; Heijnen et al., 1995; Phyle et al., 1995; Kalmbach et al., 1997). Kogure and coworkers (1979, 1984) developed a method to determine bacterial viability, termed direct viable counts (DVC). In this assay, bacterial cell suspensions are incubated with nutrients (e.g., yeast extract) in the presence of the gyrase-inhibiting antibiotic nalidixic acid. The gyrase inhibitor affects DNA replication exclusively, such that bacteria that are able to use these nutrients and are sensitive to nalidixic acid will elongate during the incubation.

In the present study, we investigated the bacterial metabolic potential during biofilm formation in drinking water by a combination of the above described in situ methods. We attempted to develop a new method to determine the potential of bacteria to react on nutrient addition with ribosome synthesis as indicator of bacterial viability, termed probe active counts (PAC).

MATERIALS AND METHODS

TECHNICAL FEATURES

A modified Robbins device was installed as upflow reactor connected to a water tap of the domestic drinking water distribution system at the Technical University Berlin as described by Kalmbach et al. (1997).

TOTAL CELL COUNTS

Cell counts of surface associated bacteria were determined according to a modification of the protocol of Hicks et al. (1992) for dual staining of samples with 4',6-diamidino-2-phenylindole (DAPI) and fluorescence-labeled rRNA probes. Staining was performed for 5 min at a final concentration of 1 μg DAPI ml^{-1} after hybridization and washing.

IN SITU HYBRIDIZATION

Fixation of samples and in situ hybridization of adherent biofilm communities were performed according to the protocol described by Manz et al. (1993).

OLIGONUCLEOTIDES

The following oligonucleotides were used in this study: (1) EUB338, complementary to a region of the 16S rRNA conserved in the domain bacteria (Amann et al., 1990); (2) Non-EUB338, complementary to EUB338, serving as a negative control for non-specific binding; (3) ALF1b, complementary to a region of the 16S rRNA characteristic for the α-subclass of Proteobacteria (Manz et al., 1992); (4) BET42a and GAM42a, oligonucleotides complementary to a region of the 23S rRNA of the β- (BET42a) and γ- (GAM42a) subclass of Proteobacteria (Manz et al., 1992); and (5) CF319a, specific for the flavobacteria-cytophaga group of the phylum cytophaga-flavobacter-bacteroides (Manz et al., 1996). The oligonucleotides were purchased from TIB MOLBIOL, Berlin, Germany, and labeled with tetramethylrhodamine-5-isothiocyanate (TRITC; Molecular Probes, Eugene, OR) or with 5(6)-carboxyfluorescein-*N*-hydrosuccinimide ester (FLUOS, Boehringer Mannheim, Germany) as described previously (Amann et al., 1990).

CTC STAINING

Biofilms were removed from the Robbins device after different exposure times and immediately placed in sterile drinking water or in R2A medium (Reasoner and Geldreich, 1985) diluted with distilled water (1:1), both containing 0.5 mM 5-cyano-2,3-ditolyl tetrazolium chloride (CTC; Polysciences Inc., Eppelheim, Germany). The biofilm was incubated at room temperature in the dark for 2 hr, rinsed in distilled water and fixed with formaldehyde solution (4% v/v). DAPI-staining was performed as described above, and the biofilm was mounted in Citifluor (Citifluor AF2; Citifluor Ltd., London, UK). CTC solutions were prepared immediately before use.

PROBE ACTIVE COUNTS (PAC)

The protocol for the determination of direct viable counts proposed by Kogure et al. (1984) was modified by use of pipemidic acid (analytical grade; Sigma, Deisenhofen, Germany) at a final concentration of 10 mg l^{-1} in 30 ml of diluted R2A medium instead of yeast extract, followed by incubation of biofilms at 21°C in the dark for 8 hr. After incubation, biofilms were fixed in 4% formaldehyde solution (v/v) for 2 hr, washed once in phosphate-buffered saline (130 mM NaCl, 7 mM Na$_2$HPO$_4$, 3 mM NaH$_2$PO$_4$, pH 7.2) and dried at 21°C. Enumeration of elongated cells was replaced by determination of probe active counts after in situ hybridization using the bacteria-specific probe EUB338 and microscopic enumeration of fluorescent cells. To determine the amount of cells detached from the biofilm during incubation, the medium was filtered through a polycarbonate membrane filter (0.2-μm pore size); subsequently, cells were stained with DAPI as described above. The number of detached bacteria per sq. cm of biofilm surface (D) was calculated by using the following equation:

$$D = N \times A_m/A_b$$

where N is the number of bacteria per sq. cm after filtration of the assay medium on the membrane, A_m is the membrane surface area and A_b is the surface area of the incubated biofilm. The performance of the gyrase inhibitors, pipemidic acid, nalidixic acid and piromidic acid (analytical grade; Sigma), was evaluated in parallel assays by comparing cell counts on biofilms incubated in 0.5 × R2A in the presence of one of the antibiotics with cell counts on the biofilm prior to incubation. Stock solutions of antibiotics were prepared at concentrations of 10 g l^{-1} in 0.05 M NaOH and stored at -20°C. Cell counts were determined after staining with DAPI as described above.

MICROSCOPY AND DOCUMENTATION

Fluorescence was detected by epifluorescence microscopy with a Zeiss Axioskop microscope (Oberkochen, Germany) fitted with a 50W high-pressure bulb and Zeiss light filter set no. 01 for DAPI (excitation 365 nm, dichroic mirror 395 nm, suppression 397 nm), no. 09 for FLUOS (excitation 450–490 nm, dichroic mirror 510 nm, suppression 520 nm) and no. 15 for CTC (excitation 546 nm, dichroic mirror 580 nm, suppression 590 nm). Color micrographs were recorded on Kodak EES 1600 color reversal film, and exposure times were 8–30 s. For statistical evaluation, at least 10 microscopic fields (100 × 100 μm) and a minimum of 1000 cells were chosen randomly and enumerated. Statistical analyses (standard error, Student's t-test) were performed with the SigmaPlot 2.0 software package (Jandel Scientific, Erkrath, Germany).

RESULTS AND DISCUSSION

IN SITU MEASUREMENT OF METABOLIC POTENTIAL

To determine changes in the metabolic potential of bacteria during the first 70 days of biofilm formation, respiratory activity by CTC reduction and fluorescence signals after in situ hybridization of single bacterial cells were monitored as independent parameters. The time course of the prokaryotic metabolic potential assessed by CTC reduction and in situ hybridization is given in Figure 1.

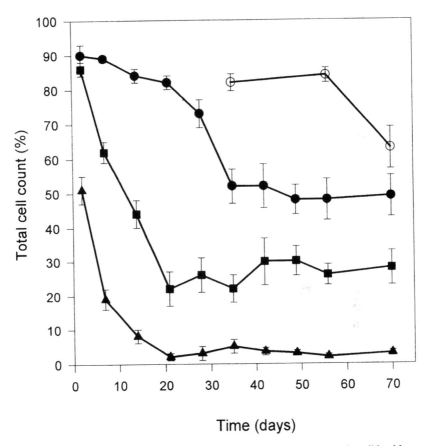

Figure 1 Changes in bacterial metabolic potential during biofilm formation on glass slides. Metabolic potential of single cells was assessed by in situ hybridization (●), respiratory activity measured by CTC reduction in sterile drinking water (◆) and CTC reduction after biofilm incubation in R2A medium (■). Probe active counts (○) were determined by in situ hybridization after incubation in R2A medium amended with pipemidic acid. Values were normalized over total cell counts by DAPI staining. Error bars represent standard errors of mean values.

In situ and Substrate Dependent CTC Reduction

Bacterial respiratory activity was visualized by the intracellular reduction of CTC to the red fluorescent CTC-formazan crystal (Rodriguez et al., 1992), a method that is widely used to enumerate respiring cells in different environmental samples (Schaule et al., 1993; Heijnen et al., 1995; Kalmbach et al., 1997). In our system, a concentration of 0.5 mM CTC was sufficient to detect the maximum number of respiring bacteria, and an increase of the dye concentration to 5 mM did not result in higher cell numbers. However, assay conditions strongly influenced the amount of CTC-reducing bacteria. As shown in Figure 1, incubation of biofilms in sterile drinking water resulted in significantly lower percentages of actively respiring cells compared with the incubation in carbon amended medium ($P < 0.0034$). Within biofilms grown for 2 days on glass surfaces, 51% ($\pm4\%$) of total cell counts reduced CTC after incubation in sterile drinking water. A significant increase of actively respiring cells to 86% ($\pm2\%$) was reached by incubation in 0.5 × R2A medium (Figure 1). The vast difference between the amount of respiring bacteria under in situ conditions and after incubation with nutrients, indicated that a remarkable fraction of cells reduced CTC only in the presence of nutrients. Although Rodriguez et al. (1992) and Smith et al. (1994) described the same effect, other studies reported incubation without nutrients to be as effective (Coallier et al., 1994; Phyle et al., 1995). These discrepancies might be due to differences in the affinity of bacterial cells obtained from various habitats to CTC and nutrients, and to variations of the incubation protocol (Phyle et al., 1995).

Decrease of Respiratory Activity during Biofilm Formation

The number of formazan-containing fluorescent cells declined continuously during the first 21 days of slide exposure and stabilized thereafter. On glass slides, the amount of in situ detectable, respiring cells after incubation in sterile drinking water reached a plateau of 2% ($\pm1\%$) to 5% ($\pm2\%$) of total cell counts, whereas the number of respiring cells after activation in 0.5 × R2A medium remained between 22% ($\pm5\%$) and 30% ($\pm7\%$) of total cell counts (Figure 1). CTC reduction by biofilm bacteria on polyethylene as substratum was monitored in a parallel experiment. The amount of CTC-reducing biofilm bacteria was comparable with the percentage of respiring cells on glass surfaces during the experiment. The number of respiring cells stabilized at 3% in drinking water and at 30% after incubation in 0.5 × R2A, respectively.

Assessment of Bacterial Ribosome Content

Early studies on the correlation of bacterial ribosome content and growth rate of *Salmonella typhimurium* (Schaechter et al., 1958) have been adopted for conclusions about bacterial growth rates by measuring fluorescence intensities after in situ whole-cell hybridizations (Poulsen et al., 1993; Møller et al., 1995). Poulsen and coworkers (1993) also demonstrated that decreasing hybridization signal intensities were correlated with lower rRNA contents of the bacteria and are not caused by reduced cell permeability. The proportion of bacteria containing a sufficient amount of ribosomes to yield positive hybridization signals was determined by using the bacteria-specific oligonucleotide probe EUB338.

Ribosome Content and Percentage of CTC-Reducing Cells during Biofilm Formation

In contrast to the number of CTC-reducing cells, the amount of cells with positive hybridization signals declined very slowly during the first 21 days of biofilm formation on glass surfaces from 90% ($\pm$3%) to 82% ($\pm$2%) (Figure 1). After 35 days of exposure, the percentage of detectable cells after hybridization was reduced to 52% ($\pm$5%) of total cell counts and remained stable at this level. This decrease of fluorescence signal intensity was observed in all experiments performed on biofilms on glass and polyethylene slides. Comparative studies between these two independent methods showed that fluorescence signals and the supposed corresponding cellular ribosome content declined more slowly than bacterial in situ or substrate-dependent respiratory activity. These findings are in agreement with results from Fukui et al. (1996), who suggest a certain amount of a durable fraction of rRNAs in starved bacterial cells.

PROBE ACTIVE COUNTS (PAC)

In drinking water, as in most natural and man-made habitats, more than 90% of the total bacterial population could not be enumerated by heterotrophic plate counts, because unfavorable culture conditions or the formation of viable but non-culturable states (Morita, 1985). In mature drinking water biofilms, about 50% of single bacterial cells could be easily detected by in situ hybridization using fluorescence-monolabeled oligonucleotide probes. To obtain information about the activity and viability of the remaining 50% of the microbial population, a new method for the assessment of these apparently inactive bacteria was established.

Evaluation of Cell-Division Inhibiting Antibiotics

The direct viable count method (DVC) developed by Kogure et al. (1979) determines the capacity of single bacterial cells to react on nutrient addition with protein synthesis. Most surveys of bacterial direct viable counts (Byrd et al., 1991; Coallier et al., 1994; Heijnen et al., 1995) are based on the original protocol of Kogure et al. (1979), involving incubation of bacteria with yeast extract in the presence of the cell division-inhibiting antibiotic nalidixic acid and subsequent determination of elongated cells. In our preliminary experiments with drinking water biofilms, however, nalidixic acid could not effectively inhibit cell division, nor was yeast extract found to be the most suitable substrate for cell activation. The incubation protocol, therefore, was modified, and optimal conditions for investigations of drinking water biofilms were determined.

In 1984, Kogure et al. proposed the combined use of three antibiotics, nalidixic acid, piromidic acid and pipemidic acid, but they could not report a significant increase in the efficacy compared with the incubation with nalidixic as sole inhibitor. In this study, however, we could show that pipemidic acid was most effective for inhibition of cell division during the 8-hr incubation period. In a comparative analysis of nalidixic acid, pipemidic acid and piromidic acid for their efficacy within drinking water biofilms, the initial cell density of 1.2×10^6 ($\pm 9 \times 10^4$) only remained stable with pipemidic acid as gyrase inhibitor (1.1×10^6, $\pm 1 \times 10^5$, $P = 0.44$). Nalidixic acid and piromidic acid did not effectively suppress cell division, and cell numbers increased significantly from 1.2×10^6 ($\pm 9 \times 10^4$) to 2.2×10^6 ($\pm 1.8 \times 10^5$) ($P < 0.001$) and 2.8×10^6 ($\pm 1.7 \times 10^5$) ($P < 0.001$), respectively.

Although all three antibiotics inhibit DNA synthesis and prevent cell division in a similar manner, differences in their effectiveness against certain bacterial species (Kogure et al., 1984) and a wide range of marine bacterial isolates (Joux and LeBaron, 1997) have been reported. In other habitats, however, pipemidic acid might not be sufficient as the sole cell division inhibitor. Joux and LeBaron (1997) proposed the use of an antibiotic cocktail, because high numbers of marine bacterial isolates were found to be resistant to one or several gyrase inhibitors. These findings clearly demonstrate the necessity to optimize the DVC procedure for each habitat and to monitor cell numbers during the activation procedure.

Measurement of Metabolic Activity by Hybridization Signals

Enumeration of elongated cells, as proposed in the DVC method, proved to be quite difficult in mature biofilms with a morphologically heterogeneous microbial population. Therefore, enumeration was replaced by fluorescent cell counting after in situ hybridization, based on the correlation between bacterial ribosome content, growth rate and hybridization signal intensity (Schaechter et al., 1958; Poulsen et al., 1993). The amount of bacteria with a sufficient ribosome content to yield a clear fluorescence signal after in situ hybridization with the bacteria-specific oligonucleotide probe EUB338 was defined as the number of probe active counts (PAC). For evaluation of non-specific staining and autofluorescent cells, control hybridizations with the labeled non-EUB338 probe were performed.

Influence of Pipemidic Acid Concentrations

The influence of different concentrations of pipemidic acid was tested for biofilms incubated in R2A medium for incubation periods of 8, 16 and 24 hr. Concentrations of 10 μg ml^{-1} were sufficient to suppress cell division for incubation periods of 8 hr, but not for 16 or 24 hr. Higher concentrations of pipemidic acid (30 and 50 μg ml^{-1}) effectively hindered cell growth for 8 and 16 hr of incubation, but failed at longer incubation periods. The amount of bacteria yielding a clear hybridization signal after incubation with 50 μg ml^{-1} of pipemidic acid, however, varied strongly between the experiments and was, in most cases, lower compared with incubations with 30 or 10 μg ml^{-1} of the gyrase inhibitor (Figure 2).

The performance of pipemidic acid was monitored in all activation experiments by enumeration of total cell counts of surface-associated bacteria with the fluorochrome DAPI before and after activation. In addition, the mean values of cells detached from the examined biofilm during incubation were determined by microfiltration of the assay medium and staining of cells with DAPI. Total cell numbers did not change significantly in any of the activation experiments ($P > 0.18$), clearly indicating the effective suppression of bacterial cell division by pipemidic acid (Table 1). The amount of detached cells during activation varied between 2 and 6% of total bacterial counts (surface associated and detached cells).

Comparison of Different Incubation Media

R2A medium was developed for cultivation of bacteria from potable water (Reasoner and Geldreich, 1985) and, therefore, seemed to be most suitable for our experiments. A comparison between incubation of biofilms in 0.5 × R2A medium and in yeast extract, as proposed by Kogure et al. (1979, 1984), revealed a slight difference in the percentage of activated bacteria between the two media. Incubation in yeast extract resulted in 79% ($\pm$2%) probe active counts, whereas incubation in R2A resulted in 84% ($\pm$1%, $P = 0.015$) activated cells.

Results of Activation Experiments

Results of biofilm activation experiments after 35, 56 and 70 days of slide exposure are summarized in Table 1 (glass) and Table 2 (polyethylene). In general, the amount of cells yielding a bright hybridization signal with the probe EUB338 increased from about 50% to more than 80% in activation

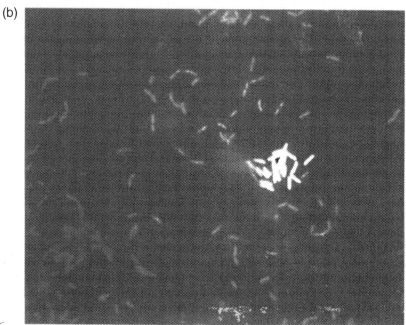

Figure 2 Epifluorescence photomicrographs of biofilm communities on polyethylene surfaces. In situ hybridizations were performed with the fluorescein-labeled probe BET42a specific for the β-subclass and the tetramethylrhodamine-labeled probe GAM42a for the γ-subclass of *Proteobacteria*. (a) Weak fluorescence signals after hybridization of a 28-day-old native biofilm. (b) Cell elongation and strong fluorescence signals of cells in a 28-day-old biofilm after 8 hr of incubation in R2A amended with pipemidic acid.

TABLE 1. Activation of Attached Bacteria on Glass Surfaces.

Detection Method	35 Days of Exposure		56 days of Exposure		70 Days of Exposure	
	Before Activation	After Activation	Before Activation	After Activation	Before Activation	After Activation
EUB338[a]	52%	82%	48%	84%	49%	63%
ALF1b[a]	11%	18%	5%	15%	3%	4%
BET42a[a]	34%	55%	36%	55%	34%	56%
GAM42a[a]	4%	4%	2%	7%	6%	4%
CF319a[a]	0	0	<1%	5%	0%	1%

[a]All values are percent of total cell counts determined by DAPI staining.

117

TABLE 2. Activation of Attached Bacteria on Polyethylene Surfaces.

Oligonucleotide Probe	35 Days of Exposure		56 days of Exposure		70 Days of Exposure	
	Before Activation	After Activation	Before Activation	After Activation	Before Activation	After Activation
EUB338[a]	58%	84%	48%	80%	52%	80%
ALF1b[a]	<1%	1%	7%	13%	4%	26%
BET42a[a]	57%	80%	37%	55%	34%	45%
GAM42a[a]	<1%	1%	2%	3%	4%	2%
CF319a[a]	0	0	1%	7%	0%	3%

[a]All values are percent of total cell counts determined by DAPI staining.

experiments, with the only exception of a 70-day-old biofilm community grown on glass, which showed a comparably low activation rate with 63% hybridized cells of total cell counts. This clearly demonstrated the rapid, subtrate-dependent reactivation of apparently inactive, adherent bacteria by addition of appropriate nutrients suitable for oligotrophic freshwater habitats.

Group-specific oligonucleotide probes were used to characterize and compare the bacterial population composition of biofilms before and after activation. The relative abundance of bacteria belonging to the β- or γ-subclass of Proteobacteria did not differ significantly between activated and non-activated biofilms on glass or polyethylene slides. Bacteria belonging to the flavobacteria-cytophaga group were present at very low numbers in mature, non-activated biofilms. After activation, however, 1–7% of the total bacterial population of the biofilm could be identified as members of this taxon. The percentage of bacteria belonging to the α-subclass of Proteobacteria increased in some biofilm populations drastically after activation (e.g., from 4 to 26% in a 70-day-old biofilm on polyethylene and from 5 to 15% in a 56-day-old biofilm on glass), but it remained at the same level in other biofilms. This is a strong indication for the presence of subpopulations of mostly inactive bacteria that might be present at some stages of the biofilm development. Bacteria belonging to the flavobacteria-cytophaga group and to the α-subclass of Proteobacteria were seriously underestimated by in situ hybridization in some mature biofilms but could be successfully visualized by performing the PAC-method. This new method, therefore, might be an important improvement for rRNA-targeted in situ probing of bacteria in extremely oligotrophic habitats.

REFERENCES

Amann, R.I., W. Ludwig and K.-H. Schleifer. 1995. Phylogenetic identification and in situ detection of individual microbial cells without cultivation. *Microbiol. Rev.* 59, 143–169.

Amann, R.I., B.J. Binder, R.J. Olson, S.W. Chisholm, R. Devereux and D.A. Stahl. 1990. Combination of 16S rRNA-targeted oligonucleotide probes with flow cytometry for analyzing mixed microbial populations. *Appl. Environ. Microbiol.* 56, 1919–1925.

Byrd, J.J., H.-S. Xu and R.R. Colwell. 1991. Viable but nonculturable bacteria in drinking water. *Appl. Environ. Microbiol.* 57, 875–878.

Coallier, J., M. Prevost and A. Rompre. 1994. The optimization and application of two direct viable count methods for bacteria in distributed drinking water. *Can. J. Microbiol.* 40, 830–836.

Costerton, J.W., J.C. Nickel and T.I. Ladd. 1986. Suitable methods for the comparative study of free-living and surface-associated bacterial populations. In: J.S. Poindexter and E.R. Leadbetter (eds.): *Bacteria in nature—methods and special applications in bacterial ecology, Vol. 2,* Plenum Press, New York, 49–84.

DeLong, E.F., G.S. Wickham and N.R. Pace. 1989. Phylogenetic stains: ribosomal RNA based probes for the identification of single cells. *Science* 243, 1360–1363.

Fukui, M., Y. Suwa and Y. Urushigawa. 1996. High survival efficiency and ribosomal RNA decaying pattern of *Desulfobacter latus,* a highly specific acetate-utilizing organism, during starvation. *FEMS Microbiol. Ecol.* 19, 17–25.

Gonzales, J.M., J. Iriberri, L. Egea and I. Barcina. 1992. Characterization of culturability, protistan grazing, and death of enteric bacteria in aquatic ecosystems. *Appl. Environ. Microbiol.* 58, 998–1004.

Heijnen, C.E., S. Page and J.D. van Elsas. 1995. Metabolic activity of *Flavobacterium* strain P25 during starvation and after introduction into bulk soil and the rhizosphere of wheat. *FEMS Microbiol. Ecol.* 18, 129–138.

Hicks, R.E., R.I. Amann and D.A. Stahl. 1992. Dual staining of natural bacterioplankton with 4′,6-diamidino-2-phenylindole and fluorescent oligonucleotide probes targeting kingdom-level 16S rRNA sequences. *Appl. Environ. Microbiol.* 58, 2158–2163.

Joux, F. and P. LeBaron. 1997. Ecological implications of an improved direct viable count method for aquatic bacteria. *Appl. Environ. Microbiol.* 63, 3643–3647.

Kalmbach, S., W. Manz and U. Szewzyk. 1997. Dynamics of biofilm formation in drinking water: phylogenetic affiliation and metabolic potential of single cells assessed by formazan reduction and in situ hybridization. *FEMS Microbiol.* 22, 265–280.

Kogure, K., U. Simidu and N. Taga. 1979. A tentative direct microscopic method for counting living marine bacteria. *Can. J. Microbiol.* 25, 415–420.

Kogure, K., U. Simidu and N. Taga. 1984. An improved direct viable count method for aquatic bacteria. *Arch. Hydrobiol.* 102, 117–122.

LeChevallier, M.W., T.M. Babcook and R.G. Lee. 1987. Examination and characterization of distribution system biofilms. *Appl. Environ. Microbiol.* 53, 2714–2724.

Manz, W., R.I. Amann, W. Ludwig, M. Wagner and K.-H. Schleifer. 1992. Phylogenetic oligodeoxynucleotide probes for the major subclasses of Proteobacteria: problems and solutions. *Syst. Appl. Microbiol.* 15, 593–600.

Manz, W., R. Amann, W. Ludwig, M. Vancanneyt and K.-H. Schleifer. 1996. Application of a suite of 16S rRNA-specific oligonucleotide probes designed to investigate bacteria of the phylum cytophaga-flavobacter-bacteroides in the natural environment. *Microbiology* 142, 1097–1106.

Manz, W., U. Szewzyk, P. Ericsson, R. Amann, K.-H. Schleifer and T.-A. Stenström. 1993. In situ identification of bacteria in drinking water and adjoining biofilms by hybridization with 16S and 23S rRNA-directed fluorescent oligonucleotide probes. *Appl. Environ. Microbiol.* 59, 2293–2298.

McCoy, W.F., J.D. Bryers, J. Robbins and J.W. Costerton. 1981. Observations of fouling biofilm formation. *Can. J. Microbiol.* 27, 910–917.

Møller, S., C.S. Kristensen, L.K. Poulsen, J.M. Carstensen and S. Molin. 1995. Bacterial growth at surfaces: automated image analysis for quantification of growth rate-related parameters. *Appl. Environ. Microbiol.* 61, 741–748.

Morita, R.Y. 1985. Starvation and miniaturization of heterotrophs with special emphasis on the maintenance of the starved viable state. In M. Fletcher and G. Floodgate (eds.): *Bacteria in the natural environments: the effect of nutrient conditions.* Academic Press, New York, 111–130.

Pedersen, K. 1990. Biofilm development on stainless steel and PVC surfaces in drinking water. *Water Res.* 24, 239–243.

Phyle, B.H., S.C. Broadway and G.A. McFeters. 1995. Factors affecting the determination of respiratory activity on the basis of cytanoditolyl tetrazolium chloride reduction with membrane filtration. *Appl. Environ. Microbiol.* 61, 4304–4309.

Poulsen, L.K., G. Ballard and D.A. Stahl. 1993. Use of rRNA fluorescence in situ hybridization for measuring the activity of single cells in young and established biofilms. *Appl. Environ. Microbiol.* 59, 1354–1360.

Reasoner, D.J. and E.E. Geldreich. 1985. A new medium for the enumeration and subculture of bacteria from potable water. *Appl. Environ. Microbiol.* 49, 1–7.

Ridgway, H.F. and B.H. Olson. 1981. Scanning electron microscope evidence for bacterial colonization of a drinking water distribution system. *Appl. Environ. Microbiol.* 41, 274–287.

Rodriguez, G.G., D. Phipps, K. Ishiguro and H.F. Ridgway. 1992. Use of a fluorescent redox probe for direct visualization of actively respiring bacteria. *Appl. Environ. Microbiol.* 58, 1801–1808.

Roszak, D.B. and R.R. Colwell. 1987. Survival strategies of bacteria in the natural environment. *Microbiol. Rev.* 51, 365–379.

Schaechter, M., O. Maaloe and N.O. Kjeldgaard. 1958. Dependency on medium and temperature of cell size and chemical composition during balanced growth of *Salmonella typhimurium*. *J. Gen. Microbiol.* 19, 592–606.

Schaule, G., H.-C. Flemming and H.F. Ridgway. 1993. Use of 5-cyano-2,3-ditolyl tetrazolium chloride for quantifying planktonic and sessile respiring bacteria in drinking water. *Appl. Environ. Microbiol.* 59, 3850–3857.

Smith, J.J., J.P. Howington and G.A. McFeters. 1994. Survival, physiological response, and recovery of enteric bacteria exposed to a polar marine environment. *Appl. Environ. Microbiol.* 60, 2977–2984.

Staley, J.T. and A. Konopka. 1985. Measurement of in situ activities of nonphotosynthetic microorganisms in aquatic and terrestrial habitats. *Ann. Rev. Microbiol.* 39, 321–346.

CHAPTER 8

Formation and Bacterial Composition of Young, Natural Biofilms Obtained from Public Bank-Filtered Drinking Water Systems

THOMAS SCHWARTZ
SANDRA HOFFMANN
URSULA OBST

INTRODUCTION

O NE possible survival strategy of microorganisms is the colonization of solid surfaces by the formation of biofilms. Possible advantages of the bacterial adherence are the higher availability of nutrients attached to the surfaces (Marshall, 1989) and the possibility of optimal long-term settlement supported by the secretion of extracellular polymeric substances (EPS). This organic polymer matrix is described as a trap for nutrients from the water (Flemming and Geesey, 1991; Mittelman and Geesey, 1985). Generally, biofilm formation protects embedded bacteria against environmental influences such as antibiotics and other bioactive compounds (LeChevallier, 1991; Fletcher, 1991). The advantages of biofilm processes were used in water technology for treatment of wastewater and for biologically activated carbon filtration. In the past, different aspects of biofilm formation have been examined to understand the occurrence and behavior of bacteria in biofilms from drinking water (Allen et al., 1980; Ridgway and Olson, 1981). Now, enzymatic and molecular biological in situ techniques are established to complete conventional culture techniques for characterization of drinking water biofilms (Kalmbach et al., 1997a, 1997b; Schaule et al., 1993; Manz et al., 1993).

In the present study, the first stage of formation and composition of natural biofilms was observed in bank-filtered drinking water during and after conditioning on different pipeline construction materials. In Germany, drinking

Reprinted from *Water Research*, Vol. 32, No. 9, T. Schwartz, S. Hoffmann and U. Obst, "Formation and Bacterial Composition of Young, Natural Biofilms Obtained from Public Bank-Filtered Drinking Water Systems," pp. 2787–2797. Copyright © 1998, with permission from Elsevier Science.

water is rarely obtained directly from surface waters; it is prepurified by bank filtration or obtained from groundwater. At the waterworks Petersaue of the city of Mainz, which is located on an island in the river Rhine, the raw water is delivered from horizontal and deep wells. During conditioning, iron and manganese ions are oxidized and removed by sand filtration followed by a granular activated carbon filtration. The conditioned water is disinfected by chlorine dioxide and fed to the public distribution system of the city of Mainz. To study the natural biofilm formation, several sampling points have been set up at the plant and in house-branch connections within the distribution system by using devices adapted to the public pipeline system.

Recently, CTC and DAPI staining was used to quantify the percentage of respiring aerobic and facultative anaerobic bacteria in biofilm samples. CTC will be reduced by dehydrogenases of respiring bacteria, forming an insoluble, red fluorescent formazan crystal within the cell (Schaule et al., 1993; Rodriguez, 1992). DAPI binds non-specifically to all DNA molecules, inducing a blue fluorescence (McFeters et al., 1995). Fluorescence-labeled 16S and 23S rRNA-directed oligonucleotides were used for in situ hybridization to quantify the occurrence of target sequences in a mixture of nucleic acids within a biofilm population (Amann et al., 1995; Manz et al., 1992; Manz et al., 1993). The presence of pathogenic or facultative pathogenic bacteria in young biofilms has been analyzed by amplification of rRNA-related sequences by polymerase chain reaction (PCR), followed by Southern blot hybridization with oligonucleotide probes specific for *Legionella* and fecal streptococci (Manz et al., 1995; Beimfohr et al., 1993; Betzl et al., 1990).

MATERIALS AND METHODS

SAMPLING OF BIOFILMS

The installation of special devices directly in the drinking water pipelines has been fundamental for this study. The construction of the devices was based on the experiences of the group of Prof. U. Szewzyk, TU Berlin (Manz et al., 1993; Kalmbach et al., 1997a). They modified the Robbins device technique (Ruseska et al., 1982). In this study, we used devices adapted to our special requirements and ordered them from the Swedish company MEKVA AB, Mossvägen 1, 17540 Järfalla, Sweden. The main part in each of these devices consists of a hollow stainless steel cylindrical element (260 mm in length and 150 mm in diameter), where stainless steel bolts holding coupons of test materials can be screwed into place. Each coupon of test material (15 × 30 × 2.5 mm) is attached to the end of a bolt with a small screw. Therefore, no organic adhesives, which could influence the bacterial growth, were necessary to fix the coupons. Perforated plates just behind the inlet and before the outlet of the device provide equal distribution of water flow.

THE DIFFERENT SAMPLING POINTS AND MATERIALS

To study the natural formation of biofilms, modified Robbins devices were installed at different sampling points at the waterworks and within the distribution system of the city of Mainz. Two units were installed at the waterworks Petersaue downstream from the granular activated carbon (GAC) filters and the disinfection step (DIS) with a water flow-through of 6.0 ± 0.5 m^3 day^{-1}. Chlorine dioxide was added at a concentration of 0.12–0.16 mg l^{-1} according to the German drinking water regulations. The finished distribution water contained a chlorine dioxide residual of 0.05–0.11 mg l^{-1}. Two additional devices were installed in house-branch connections of the distribution system (D1 and D2). Both sampling points were supplied with bank-filtered drinking water from the waterworks Petersaue and were located at a distance of 1–2 km from the plant. At D1, where the house connection was performed with old ductile cast iron material, the water flow was 1.8 m^3 day^{-1}. The house connection at D2 was performed with polyethylene (1 year old). The water flow-through was about 2.3 ± 0.2 m^3 day^{-1}.

The following materials, which meet all standard requirements for drinking water pipelines and are widely used in drinking water facilities, were tested: (1) hardened polyethylene (PE-HD, MW > 200.000); (2) hardened polyvinyl chloride (PVC); (3) copper and (4) stainless steel.

CTC REDUCTION ASSAY AND DAPI COUNTERSTAINING

For maximum enumeration of respiring bacteria, the CTC (Polyscience, Eppelheim, Germany) solution was freshly prepared at a concentration of 5 mM in addition to pyruvic acid (2 mM). It was necessary to start the incubation of the coupons immediately after removal from the devices. After a 3-hr incubation period in darkness and at room temperature, the reduction assay was terminated. Because of the autofluorescence of the plastic materials, the biofilms were scraped from the surface by using a sterile cell scraper (Nunc, Germany), suspended and homogenized in the previous incubation solution. In previous experiments the removal efficiency was examined by microscopic analysis of DAPI-stained material coupons. The sample of the final cell suspension was filtered through a polycarbonate 0.2-μm pore-size membrane filter (25 mm in diameter; Nucleopore). The filter was covered with 0.5 ml of DAPI solution (1 μg ml^{-1}; Merck, Germany) within the filter funnel apparatus (Sartorius). The DAPI solution was removed by applying vacuum to the filter funnel apparatus; the filter was rinsed with sterile water, removed, air dried, embedded with Citifluor (Citifluor Ltd., London, UK) on a glass slide and analyzed by epifluorescence microscopy.

IN SITU HYBRIDIZATION WITH FLUORESCENTLY LABELED OLIGONUCLEOTIDE PROBES

The test coupons from the devices were removed after different exposure times, and the biofilms on the surfaces were fixed immediately with 50% v/v ethanol for 30 min. The biofilms were scraped from the surface after fixation, and the solution was filtered through a polycarbonate membrane. All oligonucleotide probes labeled at the 5′-end with Cy3 (Cyanine dye, Biological Detection Systems) were purchased from MWG-Biotech, Ebersberg, Germany. The sequences were based on the published data from Amann et al. (1995) and Manz et al. (1993). The following oligonucleotides were complementary to 16S or 23S rRNA sequences of the different subclasses of the Proteobacteria: (a) beta 42a (5′-GCCTTCCCACTTCGTTT-3′) and gamma 42a (5′-GCCTTCCCACATCGTTT-3′) are complementary to regions of the 23S rRNA; (b) alpha 1b (5′-CGTTCGYTCTGAGCCAG-3′) is complementary to a region of the 16S rRNA. The *Pseudomonas*-specific probe is complementary to regions of the 23S rRNA of *Pseudomonas* spp. (5′-GCTGGCCTAGCCTTC-3′), *Pseudomonas putida* and *Pseudomonas mendocina* 5′-GCTGGCCTAACCTC-3′. The *Legionella* spp. probe (5′-CTGGTGTTCCTTCCGATC-3′) is complementary to a region of the 16S rRNA (Manz et al., 1995). The hybridizing reaction was performed with 50-ng oligonucleotide probe within a humidified chamber at 37°C for 2 hr according to Poulsen et al. (1993). For β and γ, the formamide content of the hybridization buffer was 35%, for α and *Legionella* 20% and for *Pseudomonas*-specific probes no formamide was used. The hybridization buffer contained NaCl (0.9 M), Tris/HCl (pH 7.2, 10 mM) and SDS (0.1%). The hybridized filters were then washed with this hybridization buffer without probe for 25 min at 37°C and 25 min at 42°C. In addition, the bacteria on the filter were stained with DAPI (1 μg ml^{-1}) for 10 min, dried and embedded in Citifluor for microscopic analysis.

PCR AND SOUTHERN BLOT HYBRIDIZATION

The biofilms were scraped from the surface of the coupons and suspended in autochthonous water. Aliquots from these samples were centrifuged for 15 min at 1200 rpm. The pellet was resuspended in 10 μl of sterile water for PCR (Seradest). Two primer combinations for a non-specific amplification of 23S or 16S rRNA-related DNA fragments for Eubacteria were used for PCR. The primer and probe sequences were based on published data (Amann et al., 1995; Manz et al., 1995; Beimfohr et al., 1993). For PCR the Expand™ PCR System (Boehringer Mannheim, 1993) was used. Aliquots of the PCR solution were loaded on a 1% agarose gel and separated by electrophoresis. The DNA molecules were stained with ethidium bromide to verify the PCR and transferred to nylon membranes (Quiagen, Germany). The DNA was cross-linked with the membranes by UV irridation for 3 min. The filters were hybridized with 15 pmol of an oligonucleotide probe specific for *Legionella* spp. (5'-CTGGTGTTCCTTCCGATC-3'), fecal streptococci (*E. faecalis:* 5'-TAGGTGTTGTTAGCATTTCG-3', *E. faecium:* 5'-CACACAATCGTAACATCCTA-3' and *E. gallinarum:* 5'-CACAACTGTGTAACATCC-3') or a non-specific probe for Eubacteria as a control. All probes were labeled at the 3'-end of the sequence with digoxigenin by terminal transferase reaction (Boehringer Mannheim, Germany). Oligonucleotide probes, which hybridized with specific target DNA molecules, were visualized by a digoxigenin-specific antibody conjugated with alkaline phosphatase. The detection of the bound probe was performed by chemiluminescence reaction according to the manufacturer's instruction (Boehringer Mannheim, 1993) using CSPD as substrate. Positive and negative controls were performed in all experiments to confirm the specificity of the detection. The results concerning the molecular-biological detection of *Legionella* spp. were verified by the EnviroAmp™ *Legionella* kit (Perkin Elmer). Specific DNA sequences for *Legionella pneumophila* and non-pneumophila were amplified by PCR and hybridized with specific oligonucleotide probes immobilized on a test strip.

EPIFLUORESCENCE MICROSCOPY ENUMERATION

The dried polycarbonate filters were mounted on a microscope slide in antifading reagent Citifluor. A coverslip was placed on the top and pressed down to remove excess Citifluor. Bacteria on the filters were examined by epifluorescence microscopy with a magnification of 1000× (Axioplan; Zeiss, Oberkochen, Germany) using light filters for DAPI (BP 365, FT 395, LP 397) or CTC and Cy3 (BP 546, FT 580, LP 590). For statistical evaluation, the number of cells observed in 10 microscopic ocular grid fields per sample were counted. The number of cells per cm^2 of testing material was calculated from the average number of cells per ocular grid (0.0156 mm^2), the filter area and the filtered sample volume.

HETEROTROPHIC PLATE COUNTS

Bacteria were removed from the coupon surface with a sterile plastic scraper and pooled in a total volume of 10 ml of water. The heterotrophic plate count (HPC) of bacteria was evaluated by plating serial dilutions of the obtained bacterial suspension on R2A agar (Difco) (Reasoner and Geldreich, 1995); plate counts were enumerated after 5 days of incubation at 20°C. About 10 colonies from each sample were examined by Gram reaction and oxidase reaction.

RESULTS

TOTAL CELL COUNT AND COLONY-FORMING UNIT FROM PIPE MATERIAL BIOFILMS

All tested materials were rapidly colonized by bacteria, but the population densities varied significantly. The highest total cell count was measured on PE-HD and PVC at the granulated activated carbon (GAC) filter. The total counts were about 2.5×10^6 ($\pm 5.0 \times 10^5$) cells cm^{-2} for both materials after 9 days of incubation (Figure 1). In contrast to the plastic materials, metal coupons were colonized in lower density. Stainless steel coupons were colonized with 1.3×10^6 ($\pm 3.0 \times 10^5$) cells cm^{-2}, whereas on copper a total cell count of 5.0×10^5 ($\pm 5.0 \times 10^4$) cm^{-2} was measured after 9 days incubation at GAC. After disinfection (DIS) the total cell counts on all materials were reduced. One order of magnitude less cell counts were determined on plastic materials at the sampling point DIS (4.0×10^5 cells cm^{-2}), on steel the bacterial cell count was about 3.0×10^5 ($\pm 4.5 \times 10^4$) cells cm^{-2} and 4.2×10^4 cells cm^{-2} on copper. Despite the reduction of the bacterial cell count by chlorine dioxide, a biofilm formation within the distribution system (D1, D2) was observed during the short incubation time of 12 days (D1) and 15 days (D2). Again, there was a dependence on the used substrata. Plastic materials reaching a cell density in an average of 1.2×10^6 ($\pm 2.5 \times 10^5$) cells cm^{-2}. In comparison with PE-HD and PVC, the analysis of metal coupons at D1 and D2 showed again reduced microbial density (3.3–6.9×10^5 cells cm^{-2}).

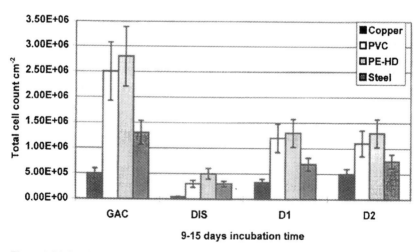

Figure 1 Biofilm formation on different pipe construction materials. The total bacterial cell count was evaluated by counting 10 microscopic ocular fields for each sample of test material in 2–3 independent experiments (GAC, granular activated carbon filtration; DIS, disinfection; D1 and D2, two independent house branch connections).

To compare the data from the total cell count with measurements of the heterotrophic plate count (HPC), bacteria from the coupons were spread on R2A plates to estimate the culturable bacterial densities on the different materials. Results of HPC determinations showed that biofilm densities of culturable bacteria on the different materials were about one order of magnitude less than those found by DAPI-staining technique (Figure 2). Obviously, the percentage of culturable bacteria was higher in biofilms from sampling point GAC than in biofilms from sampling point D1 and D2. The number of colony-forming units from plastic and steel slides was about 4.2×10^5 cfu cm^{-2} at GAC, and about 1.7×10^5 cfu cm^{-2} at D1 and D2 on average. The number of colony-forming units from copper coupons was very low at every sampling point. In contrast to the total cell count analysis, there were no significant differences concerning the colony-forming units between biofilms from steel and plastic slides. To give a rough characterization of the culturable bacteria, 250 isolates were investigated by Gram and oxidase reaction. More than 95% of them were Gram negative and had cytochrome oxidase activity.

INVESTIGATION OF THE RESPIRATORY ACTIVITY OF BIOFILMS BY CTC REDUCTION ASSAY

To investigate the respiratory activity of bacteria during the process of biofilm formation, biofilms on polyethylene (PE-HD) coupons were analyzed after different exposure times (4–36 days) at the sampling points GAC, D1 and D2. The assay for respiratory activity is based on the intracellular reduction of the tetrazolium salt CTC to red fluorescent, water-insoluble formazan crystals (Schaule et al., 1993). Previous incubation experiments with 5 mM CTC, to-

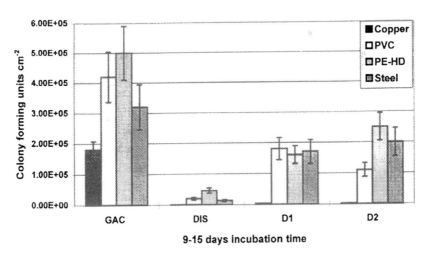

Figure 2 The colony-forming unit was evaluated by plating of biofilm suspensions on R2A agar in serial dilution.

gether with 2 mM pyruvic acid, resulted in the detection of the maximum number of respiring bacteria. The concentration of pyruvic acid is equal to the amount in the R2A medium. Other authors also observed an induction of bacterial respiration by the addition of R2A medium to the CTC incubation assay (Kalmbach et al., 1997a). The colonization of bacteria on PE-HD coupons occurred within 4 days, suggesting a very short initiation phase at the indicated sampling points (Figure 3). The highest density of microorganisms on PE-HD coupons was observed at the sampling point GAC at the water works. A total cell count of about 2.2×10^6 ($\pm 0.9 \times 10^6$) cells cm^{-2} remained relatively constant during the first 19 days of exposure time but increased to 3.8×10^6 (-0.7×10^6) cells cm^{-2} after an additional 2-week incubation within the flow device. The number of respiring cells was about 2.4×10^6 ($\pm 0.8 \times 10^6$) cells cm^{-2} after 4 days and 1.4×10^6 ($\pm 0.4 \times 10^6$) cells cm^{-2} after 7 days, respectively. Indicating that, on average. 70% of the total biofilm bacteria was metabolically active. The amount decreased to 40% in 19-day-old biofilms ($1.6 \pm 0.4 \times 10^6$ cells cm^{-2}) and increased again to 3.0×10^6 ($\pm 0.5 \times 10^6$) cells cm^{-2} (i.e., about 75%) after 34 days of incubation of PE-HD coupons at GAC.

A significant reduction of the total count of bacteria was measured at sampling point DIS as described before. Disinfection diminished the population by about 70–80% in younger biofilms and up to 50% in 19-day-old biofilms. As expected, respiratory activity was reduced by up to 80%.

In comparison with the results from the sampling point GAC, the biofilm formation occurred more slowly at the two installation points, D1 and D2, within the distribution system (Figure 3). At sampling points D1/D2 the total cell count increased slightly from 7.1×10^5 ($\pm 9.5 \times 10^4$) cells cm^{-2} after 4–6 days to 2.3×10^6 ($\pm 0.6 \times 10^6$) cells cm^{-2} after 34–36 days. In contrast to the sampling point GAC, the highest density of respiring bacteria within the biofilms at D1/D2 was quantified with 0.9×10^6 ($\pm 3.5 \times 10^5$) cells cm^{-2} (38%) after 34–36 days of incubation and the lowest amount was observed after 4–6 days of incubation with 1.5×10^5 ($\pm 2.5 \times 10^4$) cells cm^{-2} (21%). Therefore, the older the age of the biofilms, the higher the percentage of respiring bacteria at the sampling points D1/D2 within the distribution system.

To investigate the metabolic potential of biofilms from polyvinylchloride (PVC), steel and copper, slides were also incubated for 15 days at D1 and D2, respectively (Table 1). The total cell count of biofilm grown on PVC was 1.4×10^6 cells cm^{-2}, in which the percentage of respiring bacteria was about 35% (4.9×10^5 cells cm^{-2}). Steel coupons were colonized to a density of 7.3×10^5 cells cm^{-2}, in which 2.6×10^5 bacteria cm^{-2} reduced CTC to formazan crystals (35%). The total cell count in biofilms from copper slides was 5.0×10^5 cells cm^{-2}, but respiring bacteria could only be detected at a percentage below 10% in this experiment.

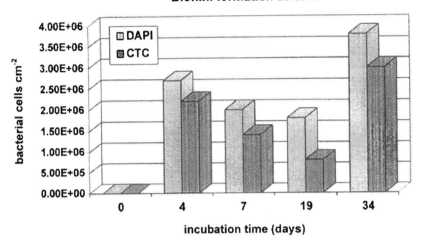

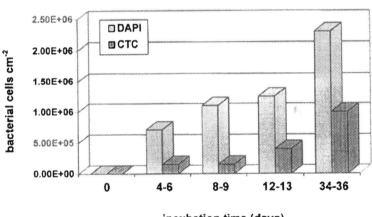

Figure 3 Total cell count and respiratory activities of biofilms on hardened polyethylene coupons (PE-HD) at the different sampling points (GAC, ganular activated carbon filtration, $n = 2$–3; D1/D2, two different house branch connections, $n = 4$–5). The results are the average of 10 microsopic ocular grid enumerations of DAPI staining (SD, 10–12%) and CTC assay (SD, 9%) of each sample.

131

TABLE 1. Total Bacterial Cell Count and Bacterial Metabolic
Potential of Biofilms from PE-HD, PVC-HD, Steel and Copper at
House Branch Connections (D1 and D2) after Incubation (15 Days)
within the Devices (n = 4–5).

Detection Method	House Branch Connections within the Distribution System			
	PE-HD	PVC-HD	Steel	Copper
Total bacterial cell count of the biofilms (DAPI)[a]	1.3×10^6 $(\pm 1.5 \times 10^5)$	1.4×10^6 $(\pm 1.7 \times 10^5)$	7.3×10^5 $(\pm 0.3 \times 10^5)$	5.0×10^5 $(\pm 0.3 \times 10^5)$
Respiration activity (CTC)[b]	38 (± 7.6)%	35 (± 6.3)%	35 (± 7.7)%	<10%

[a]Mean values from D1 and D2 (given in cells per cm^2) ± standard error.
[b]All values are percent of total cell counts determined by DAPI staining ± standard error.

IN SITU HYBRIDIZATION OF BIOFILMS GROWN IN BANK-FILTERED DRINKING WATER

The bacterial communities of biofilms from the early stage of development were analyzed by in situ hybridization with subclass-specific probes of Proteobacteria. Concerning the age of the investigated biofilms, a decrease of intensity of hybridization signals was observed in biofilms from sampling point D1, if the exposure time exceeded 7 days of incubation. In consequence, 4-day-old biofilms from GAC and D1 were compared in this study. Because of this short exposure time, the removed biofilm bacteria from two slides of each material from sampling point D1 were combined to increase the bacterial cell count for better microscopic analysis. The total cell counts of bacteria from PE-HD were 1.8×10^6 cells cm^{-2} at GAC and 6.9×10^5 cm^{-2} at D1. Similar total cell counts were measured for PVC at GAC and D1. For steel, the total amount of bacteria removed from the slide was 1.3×10^6 cells cm^{-2} at GAC and 3.2×10^5 cells cm^{-2} at D1. The percentage of the three subclasses of Proteobacteria is shown in Table 2. The plant-derived biofilms grown on plastic materials were composed of 35 ± 4.2% β and about 20 ± 0.8% γ-subclass bacteria. The α-subclass was found in a percentage of 10 ± 0.5% in biofilm grown on PE-HD and PVC. In contrast to these results, the β-subclass was the most frequent one in biofilms from steel coupons (69 ± 5.0%).

The β and γ species were the predominant subclass of Proteobacteria that colonized the plastic materials from sampling point D1, representing a house branch connection. Bacteria of the α-subclass were detected again in a much lower percentage. Similar to the hybridization results at GAC, the biofilms from steel coupons were again mainly composed of the β-subclass. The observed percentages of γ and α-subclass bacteria increased slightly at this sam-

TABLE 2. In situ Hybridization of Biofilms with Specific Probes for Proteobacteria.

Detection Method	Granular Activated Carbon Filtration GAC			House Branch Connection D1		
	PE-HD	PVC	Steel	PE-HD	PVC	Steel
Total cell count of the biofilms[a]	1.0×10^6 ($\pm 0.5 \times 10^6$)	1.0×10^6 ($\pm 0.2 \times 10^6$)	1.3×10^6 ($\pm 0.2 \times 10^6$)	6.9×10^5 ($\pm 0.8 \times 10^5$)	6.0×10^5 ($\pm 0.8 \times 10^5$)	3.2×10^5 ($\pm 0.4 \times 10^5$)
In situ hybridization[b]						
Alpha (1b)	10%	10%	8%	4%	18%	23%
Beta (42a)	37%	33%	69%	34%	33%	42%
Gamma (42a)	20%	21%	11%	38%	41%	27%

[a]Mean value (given in cells per cm^2) ± standard error.
[b]All values are percent of total cell counts determined by DAPI staining.

133

pling point. The maximum percentage of bacteria hybridizing with the *Pseudomonas* spp.-specific probe was about 10% in all 4-day-old biofilms. In addition, it was not possible to hybridize bacteria from biofilm grown on copper coupons.

DETECTION OF *LEGIONELLA* AND FECAL STREPTOCOCCI IN BIOFILMS FROM COLD DRINKING WATER SYSTEMS

To investigate the occurrence of fecal streptococci and Legionellae in biofilms, material slides were exposed for 3 weeks at GAC, D1 and D2. Aliquots from the scraped biofilms were used for PCR by using primer combinations specific for 23S or 16S rRNA-related sequences for Eubacteria. The Southern hybridizations were performed with digoxigenin-labeled oligonucleotide probes specific for *Legionella* spp., fecal streptococci (*E. faecalis, E. faecium* and *E. gallinarum*) and a non-specific probe to control the PCR and the hybridization reaction. The same specific probe used for in situ hybridization was also used for the detection of *Legionella* by Southern blot hybridization. The 3′-end of the probe was labeled with digoxigenin, whereas the 5′-end was still labeled with Cy3, which did not influence the Southern blot hybridization. Table 3 indicates where distinct signals for *Legionella* spp., fecal streptococci and Eubacteria were detected. *Legionella* spp. were found in almost all biofilms from the different sampling points, whereas fecal streptococci were only found at sampling point GAC and D2 on plastic material. The non-specific probe for Eubacteria hybridized with all samples blotted on the membrane. The appearance of *Legionella* spp. in the cold water system was verified by in situ hybridization with the Cy3-labeled probe for *Legionella* spp. For in situ hybridization, 4-day-old biofilms from D1 were removed from two slides of each material and combined for filtration on one membrane to increase the total cell count (i.e., about 1.8×10^6 cells cm^{-2}). *Legionella* spp. were detected by this method in a maximum percentage of 7.0% of the total cell count at D1. To confirm the results that *Legionella* seems to be ubiquitous in drinking water biofilms, the EnviroAmp™ *Legionella* Kit (Perkin Elmer) was used. The genus *Legionella* was identified by amplifying DNA sequences with highly specific primers complementary to conserved regions in the 5S ribosomal RNA gene and probing the resultant PCR products with probes specific for *Legionella* 5S rRNA sequences. The identification of *L. pneumophila* was linked to identifying the presence of sequences from the mip (macrophage infectivity potentiator) gene. The mip gene is conserved and specific to *L. pneumophila*. The presence of *L. pneumophila* was detected after PCR amplification with mip-specific primers by hybridization to a mip-specific probe. By using this commercial detection system non-*pneumophila* *Legionella* could be detected in all biofilm samples. No *Legionella pneumophila* were detected.

TABLE 3. Detection of *Legionella* spp. in Various Biofilm Samples by PCR-Southern blot Hybridization (PCR-SBH), In Situ Hybridization and EnviroAmp™. Fecal Streptococci and Eubacteria Were Detected by PCR-SBH. Strong Hybridization Signals Are indicated by ++, Weak Signals Are Indicated by (+) and No Signal Is Described by —. Numbers for in situ Hybridization Are the Percentages of the Total Amount of Biofilm Bacteria ($n = 2$–3; n.d. = not determined).

Probes for	Methods	Granular Activated Carbon Filtration GAC				House Branch Connections within the Distribution System D1				D2			
		PE	PVC	Cu	Steel	PE	PVC	Cu	Steel	PE	PVC	Cu	Steel
Legionella spp.	PCR-SBH	++	++	(+)	++	+	+	—	(+)	++	++	(+)	+
	In situ hybridization	n.d.	n.d.	n.d.	n.d.	2.0	7.1	0	7.0	n.d.	n.d.	n.d.	n.d.
	EnviroAmp™	++	++	+	++	++	+	+	+	++	++	(+)	++
Fecal streptococci	PCR-SBH	++	—	—	—	—	—	—	—	—	+	—	—
Eubacteria	PCR-SBH	++	++	++	++	++	++	++	++	++	++	++	++

135

DISCUSSION

Controlling the biological stability of water in distribution networks is a major concern of drinking water suppliers. Biodegradable dissolved organic carbon (BDOC) passing through the water treatment plant is one factor that contributes to biofilm formation. The release of growth-promoting compounds by certain materials coming into contact with drinking water has been described as another contribution to microbial growth in drinking water distribution systems (Schoenen and Wehse, 1988; van der Kooij, 1993). The advantages of the present study in investigating the early phase of biofilm development are: (1) the use of different kinds of pipe materials used in drinking water technology, (2) the installation of devices at crucial points within the distribution system and (3) the use of different in situ investigations in combination with conventional techniques. The results demonstrated that plastic materials were colonized in higher densities than the total bacterial cell count on steel and copper in bank-filtered water at all sampling points. The specific situation after the activated carbon filters at the waterworks may contribute to the fast and increased colonization with bacteria. The biofilms on granular carbon particles are described as a monolayer facilitating an intensive contact between microorganisms and substrates dissolved in water or fixed in carbon particle pores (Stringfellow et al., 1993; Davies and McFeters, 1988). This efficient nutrient supply may be responsible for an optimized metabolic physiology and an increased growth rate of bacteria. The high amount of metabolically active bacteria localized on released carbon fines, and also planktonic microorganisms in the effluent of the granular activated carbon filters, seem to be responsible for the rapid and dense colonization of the material coupons. In agreement with these results, Sibille et al. (1997) also measured a high amount of total fixed bacteria grown on PVC coupons in GAC water in the experiment (8.8×10^6 cells cm^{-2}). Concerning the physiology of the attached bacteria, high dehydrogenase activities were observed in biofilms sampled after the filtration steps at the plant in this study. Organic compounds loaded on carbon fines, which attach to material surfaces, could support the nutrient supply of the biofilm bacteria, but also dissolved residues of dead microorganisms from carbon filtration could play an important role in nutrient supply. The disinfection with chlorine dioxide according to the German Drinking Water Regulation (0.05–0.1 mg l^{-1}) reduced the total cell count from test materials significantly. Nevertheless, physiologically active biofilms were detected at the sampling point DIS, where the disinfectant was mixed with the conditioned water from the plant. Despite the effect of chlorine dioxide, a retarded regrowth of biofilms was observed at the two house branch connection sampling points. Morin et al. (1996) and Servais et al. (1992) showed that bacteria associated with carbon fines could be carried through the disinfection barrier without injury. Therefore, these loaded particles that attach to pipelines of the distribution system could be an important factor for biofilm regrowth. The adhesion of particles and microorganisms may also be influenced by the nature of the pipeline surfaces. Thus, chlorine dioxide induces corrosion of metallic pipe materials, increasing the release of corrosion products to the bulk water and roughening the inner surface of the pipes. Also, some bacterial species induce biological corrosion on various materials (Flemming and Geesey, 1991; Kaesche, 1990). All of these effects support the regrowth of the biofilms in the distribution system. For the present study the significance of these corrosion reactions are subordinated, because of the short exposure time of the tested pipe materials and low chloride dioxide concentrations. Similar to the biofilms developed in GAC water, the regrowth of the biofilms within the distribution system was again supported by plastic materials. The percentage of metabolically active bacteria compared with the total cell count in biofilms sampled within the distribution system was about 30–35% on average, which compares favorably with data from literature (Schaule et al., 1993). Therefore, the significant differences between the total cell count and metabolic potential of biofilms from GAC to the distribution system may result from injuries to planktonic bacteria caused by disinfection. To compare conventional plating methods with staining techniques, the results showed very clearly that the sensitivity of detection and the amount of information yielded from the DAPI/CTC assay for surpass conventional techniques. Kalmbach et al. (1997b) demonstrated that the inability of bacteria from oligotrophic water systems to form colonies on commonly used media resulted from factors other than the nonviability of these bacteria. They postulated that the investigated bacterial strains are present in distinct physiological states: (1) culturable, (2) metabolically active but incapable of undergoing the sustained cellular division required for growth on artifical medium and (3) nonculturable and metabolically inactive.

During the past years several groups investigated factors influencing biofilm formation in drinking water with different results. The studies described a significant growth-rate promoting effect of the working materials PE and PVC on planktonic bacteria (Schoenen and Wehse, 1988; van der Kooij and Veenendaal, 1993). Van der Kooij (1993) developed a method to estimate the biofilm formation potential (BFP) by the use of a type of flow-through reactor. The BFP described the biofilm density (pg ATP/cm^2) on a surface during a defined exposure time. He observed that polyethylene material supports biofilm formation in a higher degree than PVC. The results of the present study showed that there was no clear difference between the two materials using in situ techniques. The different results underline the necessity to work with more sensitive techniques to answer the questions concerning the microbial instability and the potential health risk caused by natural drinking water biofilms.

The development of in situ hybridization, with rRNA-targeted oligonucleotide probes (Amann et al., 1995), allows the identification of bacteria within their natural habitat. The application of this technique for the investi-gations of oligotrophic systems like drinking water biofilms was demonstrated by previous studies (Kalmbach et al., 1997a, b). They found that during all stages of biofilm formation on polyethylene and glass, bacteria belonging to the β-subclass of Proteobacteria dominated the drinking water biofilm com-munities. In addition, in situ probing with newly designed oligonucleotides demonstrated that most of the total bacteria (more than 80% corresponding to the β-subclass) could be identified and might be culturable on an appropriate medium (Kalmbach et al., 1997b). The present study determined the phylo-genetic composition of biofilms from the different materials exposed to drink-ing water conditioned from bank-filtered river water. Here, the β- but also the γ-subclasses of Proteobacteria were found very frequently on plastic pipe ma-terials, whereas the β-subclass was detected in a higher percentage on metal-lic coupons. The differences in group-specific composition of the biofilms from the two drinking water systems seem to result from the different source of raw water. In contrast, Kalmbach et al. (1997a) exposed material coupons in drinking water, which was conditioned from groundwater. Indeed, biofilm in situ hybridization data with groundwater-conditioned drinking water of the city of Mainz confirm this assumption. Here, the β-subclass was also found to constitute the predominant group of bacteria in all biofilms (Schwartz et al., 1997).

No significant change in subclass-specific composition of biofilms from sampling point GAC to the house branch connection was observed. This in-dicates that chlorine dioxide disinfection does not select for subclass-specific growth of biofilms within the distribution system. Some bacterial species of the β-subclass (*Leptothrix* and *Sphaerotilus*) are thought to play an important role in manganese and iron oxidation and also in microbially influenced cor-rosion (Siering and Ghiorse, 1997). Future studies will analyze the presence of members of these bacteria in biofilms of the public distribution systems of the city of Mainz. The high percentage of γ-subclass bacteria in biofilms from bank-filtered water may support the presence of hygienically relevant bacte-ria, such as *Legionella* or *Pseudomonas aeruginosa*. Previous studies demon-strated that coliforms and even (other) pathogenic bacteria, loaded on carbon fines from activated carbon filtration, could be carried through a disinfection barrier to persist in biofilms of distribution systems (Morin et al., 1996).

Polymerase chain reaction (PCR) in combination with Southern blot hy-bridizations were used for the sensitive and specific detection of facultative pathogenic bacteria in biofilms in the present investigations. The high sensi-tivity of the nonradioactive digoxigenin labeling and detection system enables the detection of single-copy, human genes (Boehringer Mannheim, 1993) and also facilitates the identification of microorganisms in mixed environmental samples like bioaerosols (Neef et al., 1995). In the present study, digoxigenin-labeled oligonucleotide probes specific for *Legionella* and fecal streptococci were used (Manz et al., 1995; Beimfohr et al., 1993). The results demonstrated that *Legionella* spp., which are facultative pathogens of humans, could be found in every investigated biofilm of a drinking water distribution system. In the literature, *Legionella* spp. are described as being adapted to warm water systems, where they multiply most effectively (Frahm and Obst, 1994; Stout et al., 1985). But Lye et al. (1997) also detected *Legionella* in significant amounts in groundwater samples and potable water supplies. Concerning the materials used, copper seems to inhibit the growth of *Legionella* spp. In agreement, copper and silver ions are described to be very effective in inactivating *Legionella* (Lin et al., 1996). In general, copper ions were found to inhibit the respiratory chain of bacteria (Domek et al., 1984), which may contribute to the low biofilm formation and metabolic potential on copper material observed in this study. Concerning its nutrient supply, *Legionella* depends on a symbiotic form of life with other bacteria and protozoa (Rogers et al., 1993). Therefore, life in the community with other microorganisms in biofilms seems to be an advantage for the growth of *Legionella*, which could lead to a percentage of about 10% of the total bacterial cell count in biofilms (Rogers et al., 1993). In agreement with these results, the present study confirmed the high percentage of *Legionella* in biofilms by in situ hybridization. The fluorescence intensities of the hybridized bacteria were not uniform. The strong signals and the weak signals were counted for microscopic quantification of *Legionella* in drinking water biofilms. In the literature decreasing hybridization signals were correlated with the rRNA content of the biofilm bacteria (Poulsen et al., 1993). Thus, the results of the hybridization experiments may also include metabolically inactive *Legionella* containing sufficient amounts of residual rRNA molecules for in situ detection.

Besides enterobacteria, fecal streptococci indicate a hygienically insufficient conditioning and distribution of disinfected water. In addition, the detection of fecal strepotococci could be a prediction for acute gastrointestinal diseases (Zmirou et al., 1987; Wiedenmann et al., 1988). The rare detection of fecal streptococci within the distribution system of Mainz demonstrated that the subsoil passage of the river water and the conditioning of the raw water at the waterworks were sufficient to eliminate fecal contamination originating from the Rhine water.

In conclusion, this study confirmed results from literature that biofilm formation is supported by plastic pipeline materials and that opportunistic pathogens could persist within these biofilms. In contrast to most previous investigations, this study used no bench scale units, but analyzed the bacterial formation directly at crucial points in the drinking water distribution network. Moreover, different sensitive methods were combined in this study to get more detailed information about biofilm development and phylogenetic composition. Additional long-term exposure experiments will determine if the differences in material-related biofilm growth are a common and not a time-dependent phenomenon in biofilm formation. Reflecting the German statistics about the increasing use of synthetic materials in drinking water systems (Herz, 1996), it seems possible that the present disinfection measures will not be able to meet future requirements.

ACKNOWLEDGEMENTS

We thank Prof. Ulrich Szewzyk for help in procuring the devices from Sweden, Prof. Rudolf Amann for introducing us to the in situ hybridization technique and the technical engineers of Stadtwerke Mainz AG for installing the modified Robbins devices. We also thank Dr. Th. Rothman and Dr. M. Wiegand-Rosinus for their critical reading of the manuscript. This research was funded by the BMBF (02-WT9538/7), DVGW and Stadtwerke Mainz AG.

REFERENCES

Allen, M.J., R.H.Taylor and E. Geldreich. 1980. The occurrence of microorganisms in water main encrustations. *J. Am. Water Works Assoc.* 72, 614–625.

Amann, R.J., W. Ludwig and K.H. Schleifer. 1995. Phylogenetic identification and in situ detection of individual microbial cells without cultivation. *Microbiol. Rev.* 59, 143–169.

Beimfohr, C., A. Krause, R. Amann and K.H. Schleifer. 1993. Identification of lactococci, enterococci, and streptococci. *Syst. Appl. Microbiol.* 16, 450–456.

Betzl, D., W. Ludwig and K.H. Schleifer. 1990. Identification of lactococci and enterococci by colony hybridization with 23S rRNA-targeted oligonucleotide probes. *Appl. Environ. Microbiol.* 56, 2927–2929.

Boehringer Mannheim. 1993. The DIG system user's guide for filter hybridization. Boehringer Mannheim GmbH, Biochemica, Mannheim.

Davies, D. and G.A. McFeters. 1988. Growth and comparative physiology of *Klebsiella oxytoca* attached to granular activated carbon particles and in liquid media. *Microbial Ecology* 15, 165–175.

Domek, M.J., M.W. LeChevallier, S.C. Cameron and G.A. McFeters. 1984. Evidence for the role of copper in the injury process of coliform bacteria in drinking water. *Appl. Environ. Microbiol.* 48, 289–293.

Flemming, C.H. and G.G. Geesey. 1991. *Biofouling and biocorrosion in industrial water systems.* Springer-Verlag, Berlin, Heidelberg.

Fletcher, M.M. 1991. The physiological activity of bacteria attached to solid surfaces. *Adv. Microb. Physiol.* 32, 53–85.

Frahm, E. and U. Obst. 1994. Experiences with improved methods for the detection of *Legionellae* in drinking water. *Zbl. Hyg.* 196, 170–180.

Herz, R.K. 1996. Ageing process and rehabilitation needs of drinking water distribution networks. *J. Water SRT-Aqua.* 45, 221–231.

Kaesche, H. 1990. *The corrosion of metals.* Springer-Verlag, Berlin, Heidelberg.

Kalmbach, S., W. Manz and U. Szewzyk. 1997a. Dynamics of biofilm formation in drinking water: phylogenetic affiliation and metabolic potential of single cells assessed by formazan reduction and in situ hybridization. *FEMS Microbiol. Ecol.* 22, 265–279.

Kalmbach, S., W. Manz and Szewzyk. 1997b. Isolation of new bacterial species from drinking water biofilms and proof of their in situ dominance with highly specific 16S rRNA probes. *Appl. Environ. Microbiol.* 63, 4164–4170.

LeChevallier, M.W. 1991. Biocides and the current status of biofouling control in water systems. In: H.C. Flemming and G.G. Geesey (eds.): *Biofouling and biocorrosion in industrial water systems.* Springer-Verlag, Berlin, Heidelberg, 113–154.

Lin, Y.-S.E., R.D. Vidic, J.E. Stout and V. Yu. 1996. Individual and combined effects of copper and silver ions on inactivation of *Legionella pneumophila. Water Res.* 30, 1905–1913.

Lye, D., S. Fout, S.R. Crout, R. Danielson, C.L. Thio and C.M. Paszko-Kolva. 1997. Survey of ground, surface, and potable waters for the presence of *Legionella* species by EnviroAmp® PCR *Legionella* kit, culture, and immunofluorescent staining. *Water Res.* 31, 287–293.

Manz, W., R. Amann, W. Ludwig, M. Wagner and K.H. Schleifer. 1992. Phylogenetic oligonucleotide probes for the major subclasses of proteobacteria: problems and solutions. *Syst. Appl. Microbiol.* 15, 593–600.

Manz, W., U. Szewzyk, P. Ericsson, R. Amann, K.H. Schleifer and T.A. Stenstrom. 1993. In situ identification of bacteria in drinking water and adjoining biofilms by hybridization with 16S and 23S rRNA-directed fluorescent oligonucleotide probes. *Appl. Environ. Microbiol.* 59, 2293–2298.

Manz W., R. Amann, R. Szewzyk, U. Szewzyk, T.A. Stenström, P. Hutzler and K.H. Schleifer. 1995. In situ identification of Legionellaceae using 16S rRNA-targeted olidonucleotide probes and confocal laser scanning microscopy. *Microbiology* 141, 29–39.

Marshall, K.C. 1989. Growth of bacteria on surface-bound substrates: significance in biofilm development. In: T. Hattori, Y. Ishida, Y. Marayuma, R. Morita and A. Uchida (eds.): *Recent advances in microbial ecology.* Jap. Sci. Soc. Press, Tokyo, 146–150.

McFeters, G.A., F.P. Yu, B.H. Pyle and P.S. Stewart. 1995. Physiological assessment of bacteria using fluorochromes. *J. Microbiol. Methods* 21, 1–13.

Mittelmann, M.W. and G.G. Geesey. 1985. Copper-binding characteristics of exopolymers from a freshwater-sediment bacterium. *Appl. Environ. Microbiol.* 49, 846–851.

Morin, P., A.K. Camper, W. Jones, D. Gatel and J.C. Goldman. 1996. Colonization and disinfection of biofilms hosting coliform-colonized carbon fines. *Appl. Environ. Microbiol.* 62, 4428–4432.

Neef, A., R. Amann and K.H. Schleifer. 1995. Detection of microbial cells in aerosols using nucleic acid probes. *Appl. Environ. Microbiol.* 18, 113–122.

Poulsen, L.K., G. Ballard and D. Stahl. 1993. Use of rRNA fluorescence in situ hybridization for measuring the activity of single cells in young and established biofilms. *Appl. Environ. Microbiol.* 59, 1354–1360.

Reasoner, D J. and E.E Geldreich. 1995. A new medium for the enumeration and subculture of bacteria from potable water. *Appl. Environ. Microbiol.* 49, 1–7.

Ridgway H.F. and B.H. Olson. 1981. Scanning electron microscope evidence for bacterial colonization of a drinking water distribution system. *Appl. Environ. Microbiol.* 41, 274–287.

Rodriguez, G.G., D. Phipps, K. Ishiguro and H.F. Ridgway. 1992. Use of a fluorescent redox probe for direct visualization of actively respiring bacteria. *Appl. Environ. Microbiol.* 58, 1801–1808.

Rogers, J., P.J. Dennis, J.V. Lee and C.W. Keevil. 1993. Effects of water chemistry and temperature on the survival and growth of *Legionella pneumophila* in potable water systems. In: J.M. Barbaree, R.F. Breiman and A.P. Dufour (eds.): Legionella: *current status and emerging perspectives.* American Society of Microbiology, Washington DC, 248–250.

Ruseska, I., J. Robbins, E. Lashen and J.W. Costerton 1982. Biocide testing against corrosion-causing oilfield bacteria helps control plugging. *Oil Gas J* 253–264.

Schaule, G., H.C. Flemming and H.F. Ridgway. 1993. Use of 5-cyano-2,3-ditolyl tetrazolium chloride (CTC) for quantifying planktonic and sessile respiring bacteria in drinking water. *Appl. Environ. Microbiol.* 95, 3850–3857.

Schoenen, D. and A. Wehse. 1988. Microbial colonization of water by the materials of pipes and hoses; 1st communication: changes in colony counts. *Zbl. Bakt. Hyg. B.* 186, 108–117.

Schwartz, T., S. Hoffmann and U. Obst. 1997. The influence of drinking water conditioning on the biofilm development within a drinking water distribution system. *Vom Wasser* 89, 125–138.

Servais, P., G. Billen, P. Pouillot and M. Benezet. 1992. A pilot study of biological GAC filtration in drinking-water treatment. *J. Water SRT-Aqua* 41, 163–168.

Sibille, I., L. Mathieu, J.L. Paquin, D. Gatel and J.C. Block. 1997. Microbial characteristics of a distribution system fed with nanofiltered drinking water. *Water Res.* 31, 2318–2326.

Siering, P.L. and W.C. Ghiorse. 1997. Development and application of 16S rRNA-targeted probes for detection of iron- and manganese-oxidizing sheathed bacteria in environmental samples. *Appl. Environ. Microbiol.* 63, 644–651.

Stout, J.E., V.L. Yu and M.G. Best. 1985. Ecology of *Legionella pneumophila* within water distribution systems. *Appl. Environ. Microbiol.* 49, 221–228.

Stringfellow, W.T., K. Mallon and F.A. DiGiano. 1993. Enumerating and disinfecting bacteria associated with particles released from GAC filter-adsorbers. *J. Am. Water Works Assoc.* 85, 70–80.

van der Kooij, D. 1993. Assessment of the biofilm formation potential of synthetic materials in contact with drinking water during distribution. Presented at *AWWA WQTC,* November 7–11.

van der Kooij, D. and H.R. Veenendaal. 1993. Biofilm development on surfaces in drinking water distribution systems. In: *Proc. Int. Water Supply Congr.,* Budapest, SS1, 1–7.

Wiedenmann, A., W. Langhammer and K. Botzenhart. 1988. Enterobacteria as a criterion for the quality of raw-, drinking- and swimming-pool water. Comparative study of the occurrence of enterobacteria, *Escherichia coli,* coliforms, colony counts, fecal streptococci and *Pseudomonas aeruginosa. Zbl. Bakt. Hyg. B.* 187, 91–106.

Zmirou, D., J.P. Ferley, J.F. Collin, M. Charrel and J. Berlin. 1987. A follow-up study of gastro-intestinal diseases related to bacteriologically substandard drinking water. *Am. J. Publ. Hlth.* 77, 582–584.

CHAPTER 9

The Use of Immunological Techniques and Scanning Confocal Laser Microscopy for the Characterization of Agrobacterium Tumefaciens and Pseudomonas Fluorescens, Atrazine-Utilizing Biofilms

MARTINA HAUSNER
JOHN R. LAWRENCE
GIDEON M. WOLFAARDT
MICHAEL SCHLOTER
KLAUS-PETER SEILER
A. HARTMANN

INTRODUCTION

MICROBIAL degradation is one of the most important processes affecting the fate of pesticides in the environment. The s-triazine herbicides are usually considered less susceptible to microbial attack, but they can serve as both carbon and nitrogen sources in microbial metabolism due to their N-heterocyclic ring structure (Cook and Hütter, 1981; Cook, 1987). The mineralization of atrazine, a frequently used s-triazine, may proceed via cyanuric acid, a central s-triazine metabolite (Mandelbaum et al., 1995; Radosevich et al., 1995). In an earlier part of this study, bacterial isolates were obtained from a mixed culture, enriched from near-surface groundwater and sediment samples by using cyanuric acid as the only nitrogen source and glycerin, lactate and glucose as carbon sources (Hausner, 1997). The two isolates used in this study were identified with the API identification system as *Agrobacterium tumefaciens* (isolate AT) and *Pseudomonas fluorescens* (isolate PF). Both isolates could completely degrade cyanuric acid as the only source of nitrogen, but only *A. tumefaciens* was capable of atrazine depletion under nitrogen-limiting conditions.

Some authors (e.g., Wolfaardt et al., 1994a) have shown that biofilms can facilitate the absorption and utilization of recalcitrant xenobiotics. A better un-

143

derstanding of these processes can be accomplished with the application of non-invasive in situ methods for the characterization of biofilm bacteria. Serological methods (polyclonal or monoclonal antibodies) have been successfully used in the study of complex ecosystems, such as soil, wastewater, groundwater or sediments (summarized by Schloter et al., 1995). Antibodies have also been used in combination with differential interference contrast, phase contrast or scanning electron microscopic techniques for the detection of specific microorganisms in biofilms (e.g., Rogers and Keevil, 1992; Hunik et al., 1993; Kobayaschi et al., 1988). The intense interest in biofilms coincides closely with the application of scanning confocal laser microscopy (SCLM) and digital image analysis in microbial ecology (Caldwell et al., 1992). Because SCLM also facilitates a non-invasive approach to the visualization of biofilm structure, immunofluorescence, fluorescent probes and SCLM seem to be useful tools for the purpose of in situ detection of specific microorganisms in living, non-dehydrated bacterial biofilms. The objectives of this work were to assess the application of a fluorescent polyclonal antiserum in conjunction with SCLM for the detection of atrazine- and cyanuric acid-degrading isolates in biofilms irrigated with atrazine as the sole source of nitrogen and to compare the biofilm development by the *A. tumefaciens* (AT) isolate, either as a pure culture or in a mixed culture with the *P. fluorescens* (PF) isolate.

MATERIALS AND METHODS

CULTIVATION OF STRAINS

Isolates AT and PF were grown in Luria Bertani (LB) medium (tryptone, 10 g/l; yeast extract, 5 g/l; NaCl, 2 g/l; pH 7.0) overnight at room temperature (RT). Cells were harvested by centrifugation, washed twice with PBS (phosphate-buffered saline: NaCl, 8 g/l; KCl, 0.2 g/l; Na_2HPO_4, 1.44 g/l; NaH_2PO_4 0.2 g/l; pH 7.0) and grown to midlog phase in a mineral medium (modified from Behki and Khan, 1986). The medium (pH 7.1 ± 0.1) contained the following per 1 l of double-distilled water: KH_2PO_4, 0.4 g; K_2HPO_4, 1.6 g; $MgSO_4 \cdot 7H_2O$, 0.2 g; NaCl, 0.1 g; $CaCl_2 \cdot 2 H_2O$, 29.0 mg; trace element solution, 2.0 ml. The trace element solution was adapted from Albrecht and Okon (1980) and contained per 200 ml of double-distilled H_2O Na_2MoO_4, 200 mg; $MnSO_4 \cdot H_2O$, 235 mg; H_3BO_4, 280 mg; $ZnSO_4 \cdot 7H_2O$, 24 mg. Atrazine (Institute of Organic Industrial Chemistry, Warsaw, Poland; chemical purity, 99%) was supplied as the only source of nitrogen (20 mg/l). Glucose (5 mM), glycerine (5 mM) and lactate (5 mM) were provided as carbon sources.

INOCULATION AND MAINTENANCE OF FLOW CELLS

The atrazine-grown cultures were inoculated into a continuous-flow slide culture chamber (flow cell; Lawrence et al., 1991; Wolfaardt et al., 1994a). The flow cell consisted of a plexiglass block containing eight flow-through channels (1 mm deep, 3 mm wide, 42 mm long) sealed with a #1 cover slip (45 × 50 mm). Four channels were inoculated with isolate AT (0.3 ml/channel), and the remaining channels were inoculated with a 1:1 mixture of isolates AT and PF. The channels were connected with silicone/teflon tubing to a peristaltic pump and to a medium reservoir. The cultures were injected upstream from the flow cell by using a sterile 1-ml syringe equipped with a 22-gauge syringe tip and were allowed to reside in the channel for 1 hr after inoculation before pumping the atrazine-mineral medium through the channels at a flow rate of 150 ml/day.

POLYCLONAL ANTISERUM

A polyclonal antiserum against isolate PF was raised in a 6-month-old New Zealand female rabbit by using UV-inactivated whole cells. The crude serum was fractionated on protein A columns and further purified by antiserum adsorption with highly cross-reacting strains. The purified antiserum exhibited low cross-reactivities (<10%) with relevant strains and a detection sensitivity of 10^5 cells/ml. The antigenic determinants were characterized as protein components in the outer membrane (Hausner, 1997).

VISUALIZATION OF MICROORGANISMS IN BIOFILMS

Biofilms were labeled directly in flow cells. Because multiple parallel channels were set up at the beginning of each experiment, a new channel was labeled and sacrificed each time biofilms were observed. All solutions were injected in excess upstream from the flow cell. For immunolabeling, biofilms were blocked with a 3% bovine serum albumin (BSA) for 45 min at RT and rinsed with a PBS-BSA wash (10% PBS, 0.5% BSA, 0.01% Tween 20, in double-distilled H_2O). Purified anti-isolate PF antiserum, diluted 1:10 in the PBS-BSA wash, was added and allowed to react for 1 hr at RT. Unreacted antiserum was washed out with a wash solution. An FITC-conjugated antirabbit secondary antibody, diluted 1:50 in a wash solution, was added and allowed to react for 1 hr at RT. Excess label was rinsed out with a wash solution. The immunolabeled biofilm was counterstained with a universal nucleic acid stain (SYTO 17; Molecular Probes, Eugene, OR). A reaction time of approximately 2 min was allowed, and excess dye was washed out with growth medium.

DETERMINATION OF CELL VIABILITY

The multiple-fluor probe LIVE/DEAD® *Bac*Light™ (Molecular Probes) was used as a viability indicator. The differential stain is based on membrane integrity. The stains differ in their spectral characteristics and in their ability to penetrate healthy bacterial membranes. Bacteria with intact cell membranes stain fluorescent green, whereas bacteria with damaged membranes stain fluorescent red (Molecular Probes).

ATRAZINE VISUALIZATION IN PURE CULTURE BIOFILMS

Biofilms were blocked with a 3% mouse serum blocking solution (45 min, RT) and rinsed with wash solution. Anti-atrazine antiserum diluted 1:10 in wash solution was injected and allowed to react for 1 hr at RT. Unreacted antiserum was rinsed out with a wash solution. FITC-conjugated antirabbit secondary antibody, diluted 1:50 in wash solution, was added and allowed to react for 1 hr at RT. Excess label was removed with wash solution. The immunolabeled biofilm was counterstained with SYTO 17.

LASER MICROSCOPY AND IMAGE ACQUISITION

Biofilms were observed non-destructively in flow cells with a scanning confocal laser microscope (MRC-600; Bio-Rad Microscience, Mississauga, Ontario, Canada), mounted on a Nikon-Microphot-SA microscope. The microscope was equipped with an argon laser (488 and 514 nm) and a 60×1.4 numerical aperture oil immersion lens (Nikon Corp., Chiyoda-ku, Tokyo). The microscope was operated as described by Wolfaardt et al. (1994a). Single *xy*-images from a selected optical plane, a series of optical planes (i.e., a *z*-series), and sagittal sections in the vertical (*xz*) axis were collected. Images were acquired either directly or by using Kalman filtration. Both red (SYTO 17) and green (FITC) emitted fluorescence signals were collected.

IMAGE ANALYSIS

Images were processed with a Northgate 80486 host computer and software provided by Bio-Rad, as described by Lawrence et al. (1991) and Wolfaardt et al. (1994a). Biofilm thickness was estimated from sagittal *xz*-sections. Percent cell area (cell density), population dynamics and viability were assessed for biofilm depth from *xy*-thin sections by using image thresholding. For each biofilm and parameter, five *z*-series (consisting of single *xy* scans taken at 1-μm intervals) or sagittal (*xz*-scans) were analyzed and averaged to obtain quantitative information. A more detailed discussion of these methods can be found in Caldwell et al. (1992) and Lawrence et al. (1996).

RESULTS

HORIZONTAL OPTICAL THIN SECTIONING OF BACTERIAL BIOFILMS

Optical sectioning demonstrated that both the pure (AT) and mixed culture (AT, PF) biofilms followed a similar course of development (Figure 1). In both cases, the total cell density was highest on day 7 [Figure 1(a), (d)], decreased on day 14 [Figure 1(b), (e)] and increased again on day 21 [Figure 1(c), (f)]. Both mixed and pure culture biofilms exhibited a pyramidal shape, with the highest cell density at and near the attachment surface (Figure 2), with the number of cells declining toward the outer edges of the biofilm. The pure culture biofilm contained less cellular material than the mixed culture biofilm [Figure 1, Figure 2(c), (d)].

SAGITTAL SECTIONING

Sagittal sections showed that the pure culture biofilm was diffuse with a thickness of approximately 5 μm throughout the experiment [Figure 1(g), (h), (i), Figure 2(d)]. In contrast, the mixed culture biofilm was initially more compact, with a higher cell density at the attachment surface. On day 21, saggittal sections revealed an irregular and diffuse biofilm with a disseminated foundation and vertical development in the form of outcrops of cells up to 15 μm, projecting perpendicularly into the bulk aqueous phase (Figure 1). The biofilm thickness developed from 5 μm on day 7 to 13 μm on day 21. Horizontal sectioning of the mixed culture biofilm showed that the anti-isolate PF polyclonal antibody fully penetrated the 5- to 15-μm-thick layers and labeled cells evenly at all depths of the biofilm. Isolate PF cells dominated throughout the experiment in the mixed culture biofilm. Isolate AT cells always comprised the smaller fraction of the total cellular material at all developmental stages [Figure 1(d), (e), (f)]. The cell density of isolate AT was higher in the pure culture biofilm compared with its density in the mixed culture biofilm [Figure 1(a)–(f)]. The difference was most pronounced on day 21: only a few isolate AT cells were seen in the mixed culture biofilm, whereas the pure culture biofilm was densely populated [Figure 1(c), (f)].

VISUALIZATION OF ATRAZINE

On day 21 the biofilms were labeled with a polyclonal antibody against atrazine. A particular staining pattern was revealed in the pure culture biofilm: the antibody bound to the entire attachment surface, where atrazine accumulated, except for eliptical zones of clearance, where no immunofluorescence

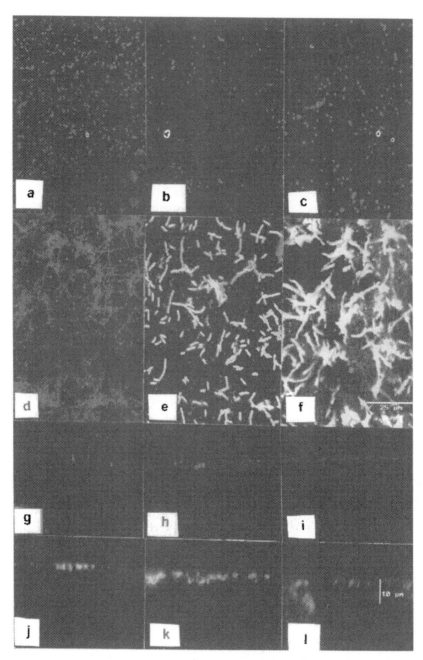

Figure 1 Horizontal [(a)–(f)] and sagittal [(g)–(l)] sections of pure [(a)–(c), (g)–(i)] and mixed [(d)–(f), (j)–(l)] culture biofilms on day 7 [(a), (d), (g), (j)], 14 [(b), (e), (h), (k)] and 21 [(c), (f), (i), (l)]. The horizontal profiles are *z*-series projections of *xy*-optical sections obtained at 1-μm intervals and collected to a depth of 5 μm. In the sagittal sections, the attachment surface is at the top of the figures. Isolate AT cells were stained with SYTO-17 (red fluorescence); isolate PF cells were labeled with FITC-conjugated antiserum (green/yellowish-green fluorescence).

148

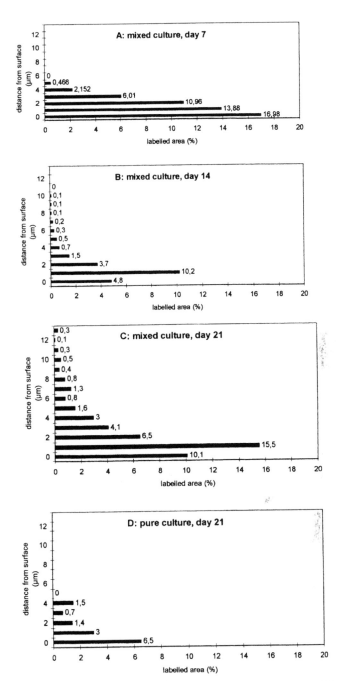

Figure 2 Quantitative analysis of horizontal sections of mixed (isolate AT/isolate PF) and pure (isolate AT) culture biofilms. The labeled cell mass (shown as percentage of the total biofilm area at different depths) is depicted at 1-week intervals (mixed culture); the distribution on day 21 only is shown for the pure culture biofilm.

was observed. After counterstaining with SYTO-17, stained isolate AT cells were found in the non-labeled zones (Figure 3).

ESTIMATION OF CELLULAR METABOLIC STATUS

The live/dead stain was used to estimate the metabolic status of the atrazine-grown pure culture biofilm on day 21 (Figure 4). It was shown that cells fluoresced predominantly green, (live); however, some cells, especially those at the attachment surface and enlarged cells near the outer edges of the biofilm, fluoresced red (dead). The live/dead stain further revealed that live cells dominated at all optical sections.

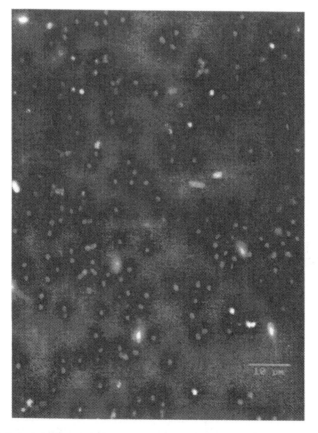

Figure 3 A single horizontal section in the *xy*-plane depicting the basal layer (0 μm) of an isolate AT pure culture biofilm on day 21. The biofilm was labeled with an FITC-conjugated anti-atrazine antiserum (green fluorescence) and counterstained with SYTO-17 nucleic acid stain (red fluorescence).

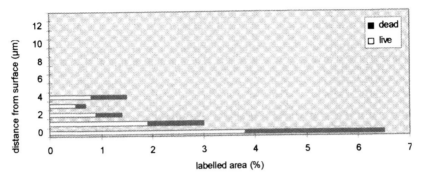

Figure 4 Quantitative analysis of horizontal sections of live/dead stained isolate AT biofilm on day 21.

DISCUSSION

An example for a successful in situ application of a fluorescence conjugated polyclonal antiserum in combination with a nucleic acid counterstain to characterize biofilms was presented. The antiserum was able to label target bacteria in biofilms ranging from 5 to 15 μm in thickness (Figure 1, Figure 2). In these biofilms, extracellular polymeric substances (EPS) likely contributing to the biofilm structure did not seem to significantly obstruct the penetration of the antiserum. Various authors have demonstrated that channel structures and pores are important features of biofilms (i.e., Lawrence et al., 1994; Massol-Deya et al., 1995). Investigating pure culture (*P. fluorescens, P. aeruginosa* or *V. para-haemolyticus*), fluorescein-stained biofilms, Lawrence et al. (1991) observed that channeling and porosity, in addition to diffusion processes, facilitated the movement of lower molecular weight molecules (e.g., 289 mol. wt. fluorescein) into the basal layers of biofilms. Lawrence et al. (1994) also demonstrated that the development of pores and channels in a diclofop-degrading mixed species biofilm allowed penetration of even high molecular weight probes (e.g. 2×10^6 mol. wt. dextrans). Because the molecular weights of IgG immunoglobulins (the major components of protein A immunoaffinity purified antisera) are also in this range (1.46×10^5 to 1.7×10^5), the success of the immunofluorescence method thus depends on the biofilm structure, especially on the amount and composition of EPS, and the existence and distribution of void spaces and channels within the biofilm. In addition, the use of BSA or another blocking agent dramatically reduces non-specific binding of antibodies to the biofilm matrix (data not shown).

Atrazine distribution was visualized in the pure culture biofilm on addition of fluorescent anti-atrazine antiserum (Figure 3). Fluorescence immunolocalization of bound atrazine residues was used in other complex systems such as plant tissue (Sohn et al., 1990) or soil (Hahn et al., 1992). Zones of clearance around isolate AT cells in the pure culture biofilm, where no bound atrazine was found, were observed following the addition of anti-atrazine polyclonal antiserum. This could be explained by bacterial use of atrazine in the vicinity of cells, possibly by diffusion of exoenzymes. Living cells were found to dominate throughout the biofilm, indicating the possibility for their participation in the atrazine metabolism (Figure 4). Some authors have suggested that the production of EPS increases under conditions of nitrogen or phosphorus limitation (e.g., Eighmy et al., 1983). Wolfaardt et al. (1994b, 1995, 1998) demonstrated the accumulation and subsequent release of diclofop methyl in optical sections of bacterial biofilms and suggested that bacterial EPS provide a mechanism for bioaccumulation of contaminants. In this regard, it would certainly be of value to investigate in detail the role of EPS produced by isolate AT in the sorption and use of atrazine.

The results presented here demonstrated that the application of polyclonal antiserum and a nucleic acid counterstain, in combination with scanning confocal laser microscopy, allowed for non-destructive in situ detection of target microorganisms (Figure 1) and for quantification of cellular distribution of biofilm depth and time (Figure 2). Accordingly, this method proved to be a useful tool for the assessment of the development of defined biofilms. Further studies are needed to assess the applicability of serological methods in natural biofilms.

ACKNOWLEDGEMENTS

This work was supported by the GSF (Research Centre for Health and Environment, Oberschleissheim, Germany) and the German-Canadian Technical and Scientific Cooperation Organization, Project KAN-ENV 55, Bioremediation of Herbicides. Special thanks to George Swerhone (National Hydrology Research Institute, Saskatoon, Canada) for his superb technical assistence.

REFERENCES

Albrecht, S. and Y. Okon. 1890. Cultures of *Azospirillum. Methods Enzymol.* 69, 740–749.

Behki R.M. and S.U. Khan. 1986. Degradation of atrazine by *Pseudomonas: N*-dealkylation and dehalogenation of atrazine and its metabolites. *J. Agric. Food Chem.* 34, 747–749.

Caldwell, D.E., D.R. Korber and J.R. Lawrence. 1992. Confocal laser microscopy and digital image analysis in microbial ecology. In: K.C. Marschall (ed.): *Advances in microbial ecology, Vol. 12.* Plenum Press, New York, 52S–66S.

Cook, A.M. 1987. Biodegradation of *s*-triazine xenobiotics. *FEMS Microbiol. Rev.* 46, 93–116.

Cook, A.M. and R. Hütter. 1981. *s*-Triazines as nitrogen sources for bacteria. *J. Agric. Food Chem.* 29, 1135–1143.

Eighmy, T.T., D. Marateca and P.L. Bishop. 1983. Electron microscopic examination of wastewater biofilm formation and structural components. *Appl. Environ. Microbiol.* 45, 1921–1931.

Hahn, A., F. Frimmel, A. Haisch, G. Henkelmann and B. Hock. 1992. Immunolabelling of atrazine residues in soil. *Z. Pflanzenernähr. Bodenk.* 155, 203–208.

Hausner, M. 1997. Studies of atrazine- and cyanuric acid-degrading bacteria in biofilms using serological methods and scanning confocal laser microscopy (SCLM). Ph.D. thesis. Ludwig-Maximilians University, Munich, Germany.

Hunik, J.H., M.P. Van den Hoogen, W. De Boer, M. Smit and J. Tramper. 1993. Quantitative determination of the spatial distribution of *Nitrosomonas europea* and *Nitrobacter agilis* cells immobilized in kappa-carrageenan gel beads by a specific fluorescent-antibody labeling technique. *Appl. Environ. Microbiol.* 59, 1951–1954.

Kobayaschi, H.A., E.C. De Macario, R.S. Williams and A.J.L. Macario. 1988. Direct characterization of methanogens in two high-rate anaerobic biological reactors. *Appl. Environ. Microbiol.* 54, 693–698.

Lawrence, J.R., G.M. Wolfaardt and D.R. Korber. 1994. Determination of diffusion coefficients in biofilms by confocal laser microscopy. *Appl. Environ. Microbiol.* 60, 1166–1173.

Lawrence, J.R., D.R. Korber, G.M. Wolfaardt and D.E. Caldwell. 1996. Analytical imaging and microscopy techniques. In: C.J. Hurst, G.R. Knudson, M.J. McInerney, L.D. Stenzenbach and M.V. Walter (eds): *Manual of environmental microbiology.* American Society for Microbiology, NY.

Lawrence, J.R., D.R. Korber, B.D. Hoyle, J.W. Costerton and D.E. Caldwell. 1991. Optical sectioning of microbial biofilms. *J. Bacteriol.* 173, 6558–6567.

Mandelbaum, R.T., D.L. Allan and L.P. Wackett. 1995. Isolation and characterization of a *Pseudomonas* sp. that mineralizes the *s*-triazine herbicide atrazine. *Appl. Environ. Microbiol.* 61, 1451–1457.

Massol-Deya, A.A., J. Whallon, R.F. Hickey and J.M. Tiedje. 1995. Channel structures in aerobic biofilms of fixed-film reactors treating contaminated groundwater. *Appl. Environ. Microbiol.* 61, 769–777.

Radosevich, M., S.J. Traina, Y.-L. Hao and O.H. Tuovinen. 1995. Degradation and mineralization of atrazine by a soil bacterial isolate. *Appl. Environ. Microbiol.* 61, 297–302.

Rogers, J. and C.W. Keevil. 1992. Immunogold and fluorescein immunolabelling of *Legionella pneumophila* within an aquatic biofilm visualized by using eoiscopic differential interference contrast microscopy. *Appl. Environ. Microbiol.* 58, 2326–2330.

Schloter M., B. Aßmus and A. Hartmann. 1995. The use of immunological methods to detect and identify bacteria in the environment. *Biotechnol. Adv.* 13, 75–90.

Sohn, G., C. Sautter and B. Hock. 1990. Fluorescence immunolocalization of bound atrazine residues in plant tissue. *Planta* 181, 199–203.

Wolfaardt, G.M., J.R. Lawrence, R.D. Robarts and D.E. Caldwell. 1995. Bioaccumulation the herbicide diclofop in extracellular polymers and its utilization by a biofilm community during starvation. *Appl. Environ. Microbiol.* 61, 152–158.

Wolfaardt, G.M., J.R. Lawrence, R.D. Robarts and D.E. Caldwell. 1998. In situ characterization of biofilm exopolymers involved in the accumulation of chlorinated organics. *Microbiol. Ecol.* 35, 213–223.

Wolfaardt, G.M., J.R. Lawrence, R D. Robarts, S.J. Caldwell and D.E. Caldwell. 1994a. Multicellular organization in a degradative biofilm community. *Appl. Environ. Microbiol.* 60, 434–446.

Wolfaardt, G.M., J.R. Lawrence, J.V. Headley, R.D. Robarts and D.E. Caldwell. 1994b. Microbial exopolymers provide a mechanism for bioaccumulation of contaminants. *Microbiol. Ecol.* 27, 279–291.

CHAPTER 10

The *in situ* Detection of a Microbial Biofilm Community on Karst Rock Coupon in a Groundwater Habitat

E. MÜLLER
B. AßMUS
A. HARTMANN
K.-P. SEILER

INTRODUCTION

IN the Southern Franconian Alb (Germany, 80 km north of Munich) the subsoil consists of two different karst rock formations: bedded and reef facies (Seiler et al., 1991). The bedded facies, a chemical sediment, has low porosity, high groundwater flow velocity (>500 m/day) and, consequently, a very low retention capacity of contaminants. The reef facies, originating from biogenic sediments, has high porosity with long duration time. In agricultural areas with high-nitrate input, the contents in the groundwater in the reef facies (2–5 mg NO_3^--N/l) were generally lower than in the bedded facies (>11 mg NO_3^--N/l) (Seiler et al., 1996). The question arises whether in the matrix of the reef facies microbiological denitrification contributes to the nitrate losses. The microbial subsurface community is composed of unattached (planktonic cells) and attached microorganisms on the sediment surface (biofilm). Unattached groundwater bacteria represent only a small proportion of the total number of cells in an aquifer (Harvey et al., 1984). Hirsch and Rades-Rohkohl (1988) found that unattached and attached groundwater communities appear to be different, although there probably are overlapping components of a dynamic community. Levels of denitrification and biodegradative activity, respectively, were lower in groundwater than in adjacent aquifer sediments (Federle et al., 1990; Smith et al., 1988). The major distinctions between sediment-bound and unattached bacteria were also documented by Kölber-Boelke et al. (1988), who performed a battery of physiological and morphological tests on bacterial isolates from water and sediment from the same borehole. Therefore, it is important to investigate the role of biofilm subsurface communities of karst rock in the microbial denitrification process.

MATERIALS AND METHODS

KARST ROCK COUPON

It is very expensive and difficult to obtain original sediment material with intact biofilm of the karst rock aquifer. For this reason, sections of rock drill cores of the reef facies (porosity: 8–15%) were cut into coupons (2 × 2 × 0.5 cm) and incubated over a period of 11 weeks in a groundwater well (area: reef facies with agriculture land use). In addition to this approach, some coupons were incubated with R_2A-medium (Reasoner and Geldreich, 1985) inoculated with *Pseudomonas fluorescens* (time of incubation: 8 weeks, 20°C).

LIVE/DEAD *BAC*LIGHT VIABILITY KIT

At the end of the incubation period some of these coupons were stained with the Live/Dead *Bac*light Viability kit (Molecular Probes Inc., Eugene, OR). One milliliter of PBS was mixed with 1.5 μl of stain 1 and 1.5 μl of stain 2 of this kit. One hundred microliters of this mixture was applied to the surface of the coupons and incubated for 3 min at 21°C. After the incubation the surface of the coupons was washed three times with H_2O_{dest}. The Live/Dead kit can detect simultaneously living and dead bacteria. The kit consists of two stains:

- stain 1, which can penetrate an energized cell membrane and fluoresces green (living cells) at blue excitation wavelength (488 nm)
- stain 2, which cannot penetrate an energized cell membrane and fluoresces red (dead cells) at blue excitation wavelength (488 nm)

IN SITU HYBRIDIZATION

Some of the incubated coupons were immediately fixed in 50% ethanol (1 hr, 20°C). After the fixation the biofilm was dehydrated by sequential washes in 80% and 100% ethanol (3 min each). Forty microliters of hybridization mixture (0.02% SDS, 20 mM Tris, 5 mM Na_3-EDTA, 0.9 M NaCl, pH 7.4), containing 250 ng of TRITC (tetramethylrhodamin-5-isothiocyanat; Moleculare Probes Inc.) labeled eubacterial probe (EUB338 according to Amann et al., 1990) was added to the surface of the coupon. The biofilm was incubated with the hybridization solution for 2 hr at 46°C in a moisture chamber. The samples were washed in a washing solution (0.02% SDS, 20 mM Tris, 5 mM Na_3-EDTA, 0.9 M NaCl, pH 7.4) for 20 min at 48°C.

CONFOCAL LASER SCANNING MICROSCOPY (CLSM)

After the live/dead staining or in situ hybridization, all samples were mounted in Citifluor (Citifluor Ltd., London, United Kingdom). All confocal images were recorded by using the 410 LSM (Carl Zeiss) mounted on an Axiovert 100 microscope equipped with 100×/1.3 and 40×/1.3 plan neofluor objectives (both with oil immersion). An argon laser with maximum emission lines at 488 and 514 nm was used as the excitation source for the live/dead stain. The fluorescence of the TRITC-labeled EUB338 probe was detected with a helium-neon laser (543 nm).

IMAGE ANALYSIS

Image processing was performed with the accompanying software package LSM software version 3.70 and the program Power Point 4.0. Reconstructed images were documented on Kodak color film (Kodak Ektachrome 100 Professional) by using an Agfa Forte slide exposure device and the program Freedom of Press.

RESULTS AND DISCUSSION

The biofilm community of the karst rock coupon surface (live/dead stain) after an 11-week incubation period in a groundwater well is demonstrated in Figures 1 and 2. The biofilm exhibited a vertical development of 1–2 layers of cells, and the biofilm thickness was approximately 10 μm. The bacterial community was composed of short rods and cocci. The colonization density of the coupon surface varied with the karst rock porosity (cf. Figure 1 and Figure 2). The results of the in situ hybridization with the labeled EUB338 probe showed very low or no TRITC fluorescence signal. The bacterial cells in the *Pseudomonas fluorescens* biofilm gave a clear fluorescent signal following the in situ hybridization with the TRITC-labeled EUB338 probe (Figure 3). This demonstrated that the setup of methods principally enabled an investigation of biofilm growing on such a rock surface.

Most microscopically visualized cells are viable but do not form visible colonies on agar plates (Amann et al., 1995). The identification of microbial populations without cultivation directly in these natural environments is very important and was the main emphasis of this experiment. The live/dead staining gave initial information about the biofilm thickness, cell density and cell morphology. The statement of Molecular Probes that this kit can detect living (green) and dead (red) cells has to be examined in further experiments. The advantage of this kit is the easy applicability and the fact that the biofilm does not need to be fixed and dehydrated. The in situ hybridization provides information about the phylogenetic composition of the microbial community and the physiological activity. If there was more knowledge about the composition of the bacterial community, the cultivation conditions could be optimized. The results of this approach demonstrate that in a groundwater habitat it is difficult to detect the bacterial cells in the biofilm by in situ hybridization with oligonucleotide probes. There are two principal reasons for the low or lacking hybridization signal in the groundwater biofilm:

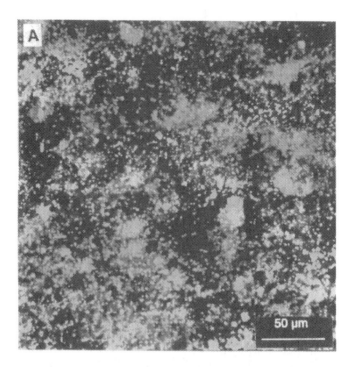

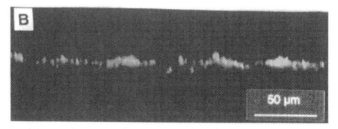

Figure 1 Biofilm on a karst rock coupon (porosity: 8%) after an 11-week incubation period in a groundwater well. Live/dead staining, green cells = living cells, red cells = dead cells. (a) LSM image: projection of a z-series (15 xy-scan, 1 µm), excitation: 488 nm. (b) LSM image: z-scan, excitation: 488 nm.

(1) The target rRNA molecule of the oligonucleotide probes is not accessible. There are two limiting diffusion barriers:

- The biofilm matrix can be a barrier to the probes; especially extracellular polymer substances (EPS) may be compacted as a result of the dehydration steps.

- In addition, the probes may not penetrate the cell wall, although membranes are expected to be readily permeable after fixation. Such

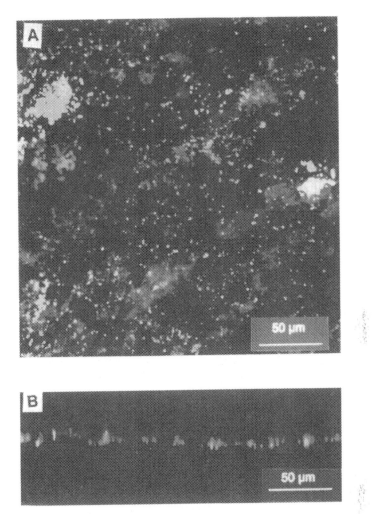

Figure 2 Biofilm on a karst rock coupon (porosity: 15%) after an 11-week incubation period in a groundwater well. Live/dead staining, green cells = living cells, red cells = dead cells. (a) LSM image: projection of a z-series (10 xy-scan, 1 μm), excitation: 488 nm. (b) LSM image: z-scan, excitation: 488 nm.

cell wall-limited probe accessibility was encountered with aldehyde-fixed Gram-positive bacteria (Hahn et al., 1992).

(2) The groundwater habitat is an oligotrophic system with slowly growing and/or physiologically less active cells. During the development of a drinking water biofilm the respiratory activity and the ribosome content of the adherent bacterial cells declined throughout the initial 35 days (Kalmbach et al., 1997). Poulsen et al. (1993) determined that the doubling rate of

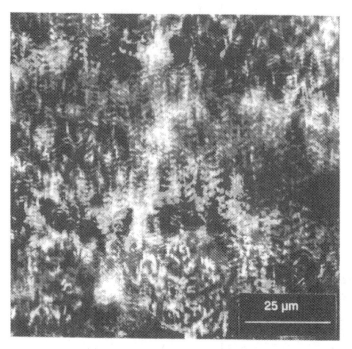

Figure 3 *Pseudomonas fluorescens* biofilm on a karst rock coupon (porosity: 8%) after an 8-week incubation period in R$_2$A-medium (20°C). *In situ* hybridization with TRITC-labeled EUB338 probe. LSM image: 3D reconstruction of a *z*-series (10 *xy*-scan, 1 μm), excitation: 543 nm.

bacterial cells in an established biofilm is significantly longer than in a young biofilm. Cells with low cellular rRNA content are difficult to detect with fluorescently labeled rRNA-targeted probes (Amann et al., 1995).

The following statements discuss the different possibilities for increasing the hybridization signals. (1) A careful treatment of the biofilm matrix, i.e., no dehydration of the biofilm, can optimize the diffusion of the probes through the EPS. However, it has to be considered that without dehydration of the biofilm the samples are not stable for a long time. (2) The use of ethanol and cell wall lytic enzymes, such as lysozyme, increases the permeability of cell walls, especially of Gram-positive bacteria. However, it is difficult to achieve a good compromise between sufficient cell permeability for efficient hybridization and good preservation of morphological details (Amann et al., 1995). (3) The detection of cells harboring low numbers of ribosomes with fluorescently mono-labeled probes might be unsuccessful. It is possible to detect whole fixed cells with oligonucleotides that are covalently linked to enzymes such as horseradish peroxidase (Amann et al., 1992). The enzyme-labeled probe is detected by the formation of an insoluble colored precipitate from suitable substrates. In addition, the sensitivity of in situ hybridization can be increased by directing multiple labeled fluorescent rRNA probes to one ribosome. The disadvantage of this approach is that the fluorescent dye molecules can attach non-specifically. (4) The use of several monolabeled oligonucleotides targeted to independent sites of the rRNA molecule resulted in more fluorescence to the target cells. Natural bacterial assemblages showed an increase of the probe-labeled fraction from 20% (1 eubacterial probe) to 75% (5 eubacterial probes) of the total population (Lee et al., 1993). The general applicability of this approach is restricted by the relatively limited availability of target sites with identical specificity. (5) The incubation of the biofilm with a supplement of different C-sources results in an increase of the cell activity. The simultaneous addition of the gyrase inhibitor nalidixic acid inhibits cell division (Kogure et al., 1979). Thus, activated cells have increased rRNA content, and the detection with oligonucleotide probes is improved. The disadvantages of the addition of C-sources are that there can be a selective stimulation of only a few organisms, which are adapted to the supplied C-substrates. This may lead to a shift in the biofilm community. It is well known that the antibiotic nalidixic acid is only effective on Gram-negative bacteria. In addition, certain Gram-negative strains are resistant to this antibiotic so that in a natural habitat there may always be some resistant bacteria. Therefore, Kogure et al. (1984) propose the use of different gyrase inhibitor antibiotic. Kalmbach et al. (1997) have shown that in a drinking water biofilm, pipemidic acid was the most effective inhibitor for cell division during an 8 hr incubation.

SUMMARY

Karst groundwater (reef facies) in an agricultural area with high-nitrate input demonstrated low-nitrate levels (2–5 mg NO_3^--N/l). This low-nitrate content may be attributed to biological degradation of nitrate in the matrix of the reef facies because of high mean residence time and microbial activity. The subsurface community consists of planktonic microorganisms and microorganisms that are living in a biofilm on the sediment subsurface. The biodegradation in the biofilm is generally higher than that in the groundwater. For the determination of the microbial denitrification in the karst subsoil, it is necessary to investigate the biofilm community. Drilled karst rock (reef facies) material was cut into coupons (2 × 2 × 0.5 cm) and incubated in a groundwater well (agricultural area) over a period of 11 weeks. After this incubation period the coupon surfaces were stained with live/dead dyes or analyzed by in situ hybridization with rRNA-directed probes. All samples were investigated by using confocal laser scanning microscopy (CLSM). The results of the live/dead stain provided information about the biofilm development. The biofilm consisted of 1–2 cell layers with a thickness of about 10 μm. The bacterial community was composed of short rods and cocci. In situ hybridization with fluorescently labeled EUB338 probe resulted in a very low or no fluorescence signal. Groundwater microorganisms, living in an oligotrophic habitat, grow slowly or may exhibit low physiological activity, which is usually correlated with low rRNA-content. Such cells are difficult to detect with oligonucleotide probes. Different possibilities for increasing the hybridization signals will be discussed.

REFERENCES

Amann, R.I., W. Ludwig and K.H. Schleifer. 1995. Phylogenetic identification and in situ detection of individual microbial cells without cultivation. *Microbiol. Rev.* 59, 143–169.

Amann, R.I., B. Zarda, D.A. Stahl and K.H. Schleifer. 1992. Identification of individual prokaryotic cells with enzyme-labeled, rRNA-targeted oligonucleotide probes. *Appl. Environ. Microbiol.* 58, 3007–3011.

Amann, R.I., B.J. Binder, R.J. Olseon, S.W. Chisholm, R. Devereux and D.A. Stahl. 1990. Combination of 16S rRNA-targeted oligonucleotide probes with flow cytometry for analyzing mixed microbial populations. *Appl. Environ. Microbiol.* 56, 1919–1925.

Federle, T.W., R.M. Ventullo and D.C. White. 1990. Spatial distribution of microbial biomass, activity, community structure, and the biodegradation of linear alkylbenzene sulfonate (LAS) and linear alcohol ethoxylate (LAE) in the subsurface. *Microbiol. Ecol.* 20, 297–313.

Hahn, D.R., R.I. Amann, W. Ludwig, A.D.L. Akkermans and K.H. Schleifer. 1992. Detection of microorganism in soil after in situ hybridization with rRNA-target, fluorescently labelled oligonucleotides. *J. Gen. Microbiol.* 138, 879–887.

Harvey, R.W., R.L. Smith and L. George. 1984. Effect of organic contamination upon microbial distributions and heterotrophic uptake in a Cape Cod, Mass., aquifer. *Appl. Environ. Microbiol.* 48, 1197–1202.

Hirsch, P. and E. Rades-Rohkohl. 1988. Some special problems in the determination of viable counts of groundwater microorganisms. *Microb. Ecol.* 16, 99–113.

Kalmbach, S., W. Manz and U. Szewzyk. 1997. Dynamics of biofilm formation in drinking water: phylogenetic affiliation and metabolic potential of single cells assessed by formazan reduction and in situ hybridization. *FEMS Microbiol. Ecol.* 22, 265–279.

Kogure, K., U. Simidu and N. Taga. 1979. A tentative direct microscopic method for counting living marine bacteria. *Can. J. Microbiol.* 25, 415–420.

Kogure, K., U. Simidu and N. Taga. 1984. An improved direct viable count method for aquatic bacteria. *Arch. Hydrobiol.* 102, 117–122.

Kölber-Boelke, J., E.-M. Anders and A. Nehrkorn. 1988. Microbial communities in the saturated groundwater environment. II. Diversity of bacterial communities in a Pleistocene sand aquifer and their *in vitro* activities. *Microbiol. Ecol.* 16, 31–48.

Lee, S., C. Malone and P.F. Kemp. 1993. Use of multiple 16S rRNA-targeted fluorescent probes to increase signal strength and measure cellular RNA from natural planktonic bacteria. *Mar. Ecol. Ptog. Ser.* 101, 193–201.

Poulsen, L.K., G. Ballard and D.A. Stahl. 1993. Use of rRNA fluorescence in situ hybridization for measuring the activity of single cells in young and established biofilms. *Appl. Environ. Microbiol.* 59, 1354–1360.

Reasoner, D.J. and E.E. Geldreich. 1985. A new medium for the enumeration and subculture of bacteria from potable water. *Appl. Environ. Microbiol.* 49, 1–7.

Seiler, K.-P., H. Behrens and H.-W. Hartmann. 1991. Das Grundwasser im Malm der Südlichen Frankenalb und Aspekte seiner Gefährdung durch anthropogene Einflüsse. *Dt. gewässerk. Mitt.* 35, 171–179.

Seiler, K.-P., E. Müller and A. Hartmann. 1996. Diffusive tracer exchanges and denitrification in the karst of southern Germany. *Proc. 4th Intern. Symp. on the Geochem. of the Earth's Crust,* 644–651.

Smith, R.L. and J.H. Duff. 1988. Denitrification in a sand and gravel aquifer. *Appl. Environ. Microbiol.* 54, 1071–1078.

CHAPTER 11

Investigation of Spatial and Temporal Gradients in Fixed-Bed Biofilm Reactors for Wastewater Treatment

AXEL WOBUS
KERSTIN RÖSKE
ISOLDE RÖSKE

INTRODUCTION

BIOLOGICAL wastewater treatment exploits the capacity of microorganisms to carry out different biochemical reactions that result in the purification of wastewater. The design of technical systems is focused on the enrichment and the maintenance of microorganisms that are able to degrade the organic substances contained in sewage and industrial effluents. One of the opportunities to achieve this objective is to separate the biomass (by sedimentation of bacterial aggregates in case of the activated sludge process) and to recycle these microorganisms into the reactor. By means of support media the microorganisms are retained in the system as a biofilm.

In such a biofilm reactor wastewater either flows over a fixed bed of support material or the biofilm-coated medium is suspended in and stirred with the liquid, i.e., the wastewater. In both cases, the purification process is based on the transport of dissolved substances into the biofilm by diffusion and sorption onto the microbial cells where they can be metabolized (Harremoës, 1978; Bryers and Characklis, 1990). Therefore, the performance of biofilm systems is determined by the mass transfer of substrates from the bulk liquid to the biofilm and within the biofilm (external and internal transport, respectively) (Harremoës, 1978; Siegrist and Gujer, 1985).

In many cases, biological fixed-bed reactors are designed with a high length/width ratio as tubes or towers characterized by *plug flow conditions*. Operating a plug flow biofilm reactor with continuous flow results in the formation of concentration gradients in the direction of flow. Therefore, it is supposed that in different zones of the reactor, biofilms with different biotic structure and different metabolic potential develop (Wilderer et al., 1993). The longitudinal heterogeneity of microbial colonization substantially affects the elimination potential of biofilm reactors. By this stratification, a high effluent quality may be achieved under constant influent conditions because of microbial communities that are well adapted to the different substrates and concentration levels. On the other hand, breakthrough events have to be expected in case of strong fluctuations of the influent concentration or of the hydraulic load. Boller and Gujer (1986) observed such a deterioration of effluent quality due to an increased load of ammonia to a tertiary nitrifying trickling filter. The displacement of substrates into sections of the filter with low biomass or with microorganisms not well adapted to this type of substrate were thought to be responsible for this effect. A similar deterioration of effluent quality was observed in the case of fixed-bed reactors treating industrial, xenobiotic-containing wastewater (Wilderer et al., 1993; Kaballo et al., 1995; Wobus and Röske, 1999).

To overcome this problem, the mode of operation has to be modified to achieve a homogeneous distribution of biomass and/or an enhanced microbial activity. Boller and Gujer (1986) suggested the operation of two filters in series with periodic inversion of the direction of flow. Long retention times and back-mixing (resulting in a flow pattern similar to completely mixed reactors) also seem to be suited for dampening strong influent fluctuations (Boller et al., 1994).

Another way to cope with a wide variety of influent conditions is to operate a reactor in a discontinuous mode that allows the adjustment of the process schedule to the actual requirements. So the *sequencing batch mode* as described by Irvine et al. (1977) has been applied to biofilm reactors (sequencing batch biofilm reactor—SBBR) by Wilderer (1992). It is to be expected thereby that longitudinal concentration gradients can be avoided and a more homogeneous spatial distribution of species and microbial activity may be achieved. In addition, periodically changing conditions are supposed to enhance the metabolic activity of microorganisms (Wilderer and Kaballo, 1995).

AIM OF THE STUDY

The objective of the experiments described in this article was to enhance the elimination potential of biofilm reactors for xenobiotics by means of special reactor design and operation strategies. Therefore, membrane biofilm reactors were constructed by using gas-permeable silicone tubings for two purposes: for oxygen supply and as support material for the biofilm (see also Wobus et al., 1995). By these means, the oxygen mass transfer was intensive enough to facilitate the enrichment of microorganisms with a high degradative potential and their maintenance in the system.

To study the influence of different operation strategies on reactor performance, we compared in our experiments two identical membrane biofilm reactors, one operated as SBBR, the other with continuous flow (continuous flow biofilm reactor—CFBR), for the elimination of chlorophenols. The chlorophenols served as model substances for poorly degradable xenobiotics of great environmental importance that have to be eliminated from industrial or municipal wastewaters. Because the reactor performance is supposed to be significantly influenced by the longitudinal distribution of biomass, its activity and the population dynamics of biofilm biocoenoses, the effects of the different modes of operation on these parameters were of central interest.

This study introduces an integrated approach for the examination of biofilm processes. Methods both for the examination of the biotic structure and for the evaluation of the metabolic potential are presented. The longitudinal heterogeneity of biofilms has to be described, and the effects on the chlorophenol elimination (also in case of fluctuation of influent conditions) has to be estimated. By analyzing biofilm dynamics due to changes of operational parameters at different levels, it becomes possible to control the performance of fixed-bed reactors.

EXPERIMENTAL SETUP

The laboratory plant (liquid volume about 10 l) has been described previously (Wobus et al., 1995). It consisted of two identical plug flow fixed-bed reactors. The helical fixed bed had 68 coils in which 6 oxygen-permeable silicone tubings were laid out in parallel. The tubings served as carrier material for the biofilm and guaranteed the oxygen supply of the microorganisms from the inner side. The membrane-grown "inverted" biofilm is characterized by opposite vertical gradients of oxygen and substrate concentration, respectively (Figure 1). To prevent compression, the tubings were sheathed by a screen of polyester wire.

The reactors were fed with the effluent from a municipal activated sludge plant to which chlorophenols at influent concentrations between 1 mg l^{-1} (2,4,5-trichlorophenol—TCP) and 80 mg l^{-1} (4-monochlorophenol—MCP) were added. The wastewater passed through the CFBR in upflow direction. To have a standard for comparison, the residence time of the water in the CFBR of 6 hr corresponded to the duration of one cycle of the SBBR, which included fill, reaction and draw periods. The SBBR was filled from the bottom to the top within half an hour. After circulating the water during the reaction period of 5 hr, the reactor was drained against the direction of filling for half an hour.

In addition to the analysis of the influent and the effluent, it was possible to take water samples at five different heights of the reactors. These sampling ports were constructed in such a way that biofilm samples could be removed after draining the reactor (also see the next section).

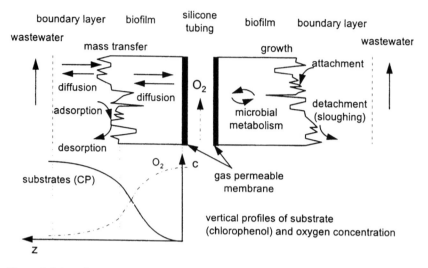

Figure 1 Schematic representation of biofilm grown on gas-permeable silicone tubings, of concentration gradients across the biofilm and of main processes significant to substrate utilization by biofilm organisms. (The thickness of the biofilm is superelevated as to the diameter of the silicone tubing.)

CHLOROPHENOL ELIMINATION BY BIOFILMS UNDER DIFFERENT OPERATING CONDITIONS

Differences in chlorophenol removal were expected between the biofilm reactors with different operation and for different sections of the CFBR. Therefore, concentration profiles along the vertical axis in the CFBR and at corresponding intervals on the timescale of the SBBR were compared. Although vertical profiles of chlorophenol concentration in the CFBR were determined by the analysis of water samples from the different sections, a composite sample (from all sampling ports) was taken from the SBBR at definite time intervals that were comparable with the duration of flow inside the CFBR.

Chlorophenol concentrations were determined by capillary gas chromatography (GC), according to the method published by Abrahamson and Xie (1983), or by high-performance liquid chromatography (HPLC). The determination of chlorophenols by gas chromatography equipped with an electron capture detector (ECD) allows a sensitive detection of less than 1 μg l^{-1} of chlorophenols with two or more Cl$^-$ substituents. Prior to the GC analysis, the chlorophenols were acetylated and extracted by the simultaneous addition of acetic anhydride and hexane to the water samples. The determination of chlorophenols by HPLC (equipped with a photodiode array detector, Waters Corp.) by using a methanol-water gradient and a C18-column (Nova-Pak®, 4 μm, 3.9 * 150 mm, Waters Corp.) is less sensitive than the former method.

However, the sensitivity of chlorophenol detection is increased to approximately 1 μg l^{-1} by solid-phase extraction (SPE) applying HR-P-columns (Macherey-Nagel GmbH) and tetrahydrofuran elution.

Elimination rates were calculated from the differences between the chlorophenol concentration at different times of a cycle. For the CFBR, the rates were determined by multiplication of concentration differences between two reactor heights with the flow rate.

As the concentration profiles in Figure 2 and the data in Table 1 show, the highest chlorophenol elimination was observed in the inflow section of the CFBR, where more than 70% of the total biomass had accumulated. According to the reduction in chlorophenol concentration, the elimination rates decreased along the vertical axis of the CFBR. In the case of higher inflow concentrations of MCP, the removal rates increased in all of the sections. In this case, the specific removal rates (per mg DNA) became similar in all sections at a load of about 70 mg l^{-1} MCP (see Table 1). However, the metabolic potential of the CFBR was obviously not high enough to eliminate this load completely. If the SBBR was subjected to such a load, an effluent concentration of more than 5 mg l^{-1} was observed subsequent to one cycle of 6 hr (Figure 2). That the effluent concentration can be kept nearly to zero by extending the reaction period has been shown in another study (Wobus and Röske, 1999).

By the comparison of elimination rates in different sections of the reactor, the heterogeneity of the chlorophenol removal potential along the direction of flow in the CFBR became evident. The highest elimination potential with regard to the surface area was observed in the zone of highest biomass. However, the relatively low removal rates per mg of DNA (i.e., per unit biomass) indicate that a considerable proportion of biomass remained inactive for

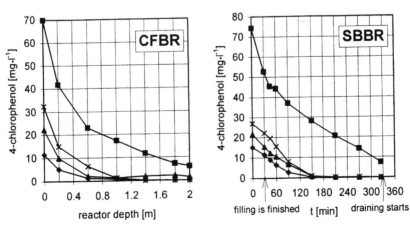

Figure 2 Profiles of the 4-chlorophenol concentration along the reactor length (CFBR) and on temporal scale (SBBR) (177th–179th day of chlorophenol addition).

TABLE 1. Comparison of Removal Rates of 4-Monochlorophenol (MCP) in Different Sections of the CFBR and of the SBBR (178th and 182nd Day of Operation). In This Experiment, the SBBR Was Divided into Three Segments by Recirculation from a Height of 0.6 m into the Bottom of the Reactor, from 1.4 to 0.6 m, and from the Top to 1.4 m Height, Respectively. In This Case, the MCP Concentrations in Reactor Heights of 0.2, 1.0 and 1.6 m, Respectively, after Filling Are Given as Influent Concentration.

Parameter	Reactor Section	CFBR: 0–0.6 m	CFBR: 0.6–1.4 m	CFBR: 1.4–2.0 m	SBBR: 0–0.6 m	SBBR: 0.6–1.4 m	SBBR: 1.4–2.0 m
Mean Biomass	DNA (mg m^{-2})	201.5	44.1	21.7	37.2	21.9	18.6
Run I							
Influent	MCP (mg l^{-1})	11.5	1.25	0.25	8.92	5.92	4.06
Effluent	MCP (mg l^{-1})	1.25	0.25	0.02	0.02	0.02	0.02
CP removal rate							
Per surface area	(mg m^{-2} h^{-1})	29.3	2.16	0.66	39.1	26.8	17.9
Per biomass (DNA content)	(mg mg^{-1} h^{-1})	0.15	0.05	0.03	1.05	1.22	0.96
Run II							
Influent	MCP (mg l^{-1})	69.9	22.9	11.7	33.3	24.0	14.9
Effluent	MCP (mg l^{-1})	22.9	11.7	5.93	0.03	0.03	0.02
Removal rate							
Per surface area	(mg m^{-2} h^{-1})	134.8	24.2	16.5	69.3	50.7	29.8
Per biomass (DNA content)	(mg mg^{-1} h^{-1})	0.67	0.55	0.76	1.86	2.32	1.60

chlorophenol degradation. In contrast, high removal rates per unit biomass were calculated for the middle and top segment of the CFBR in case of comparably influent concentrations (top section at run II and inflow section at run I, Table 1). Therefore, it is supposed that a comparably high proportion of microorganisms in this biofilm was active for chlorophenol degradation. However, the highest specific removal rates were noticed for the SBBR, which indicates a comparably high proportion of chlorophenol-degrading bacteria in this reactor. As expected, similar removal rates were determined in the different sections of the SBBR due to a more homogeneous distribution of substrate (chlorophenol) and biomass along the reactor height.

QUANTITATIVE DETERMINATION OF BIOMASS AND BIOLOGICAL ACTIVITY

Biomass quantity and the estimation of active biomass, both in general and in specific compounds, are important parameters for the control of biofilm processes (Lazarova and Manem, 1995). As the first step in quantification of biofilm biomass and biological activity, it is usually necessary to remove the biofilm from the carrier material. Direct counting of attached cells by microscopic techniques is restricted to flat surfaces and relatively thin biofilms, for example, to assess the attachment of bacteria to inert surfaces (Fletcher, 1992). The biofilm removal procedure has to meet the requirements (1) to be almost quantitative and (2) to not affect the subsequent analytical procedures.

For the investigation of membrane-grown wastewater biofilms presented in this study, 20- to 50-mm-long slices of silicone tubings were cut out after draining the reactor and removed from the reactor through the sampling ports. The biofilm was removed from the tubing by sonication in a sonic water bath and subsequent scraping (for biochemical and microbiological investigations as presented here and in the next section). Subsequently, the removed biomass was resuspended in phosphate buffer. For the examination of protozoa and metazoa, the biofilm was removed from the tubing mechanically (see the section, The Protozoa and Metazoa in Biofilms).

ESTIMATION OF BIOMASS

Various methods are available for the determination of biomass (for a review, see Lazarova and Manem, 1995). One approach is based on the direct count of (bacterial) cells by microscopic techniques. Two methods widely used for quantification of microorganisms in aquatic environments are well suited for the enumeration of microorganisms in biofilms: the detection via acridine orange (a protocol is given by Ladd and Costerton, 1990) and by DAPI (4',6-diamidino-2-phenylindol) staining introduced by Porter and Feig (1980). The latter is described in the section, Preparation of Biofilm Samples and Hybridization Protocol.

Based on cell volume and density, an estimation of bacterial biomass is possible. This represents a basic parameter in microbial ecology (Bratbak, 1993). If the measurement of length and width of individual cells is impossible, the number of 1 μm^3 is used by many authors as a rough approximation of the volume of a bacterial cell. However, this may be affected by an error of more than one order of magnitude, in particular if cell size is large due to an abundant nutrient supply. This approximation was confirmed for random samples by scanning electron microscopy. It has to be taken into consideration that microscopic methods for measuring bacterial cell size may also be inaccurate to some extent for several reasons: shrinking or swelling of cells due to fixation and dehydration, heterogeneity of microbial communities (see also Bratbak, 1993). The bacterial biomass (fresh biomass) was calculated from the biovolume of bacteria based on DAPI cell counts and an estimated density of 1.02 $\times$ 10^{-9} mg μm^{-3}. For the calculation of the biomass of eucaryotes, average values of volume and cell density have been used [to some extent available from literature (Foissner et al., 1995)].

Total biomass can also be estimated by several physicochemical or biochemical parameters obtained from the determination of specific biofilm constituents or cellular components. The measurement of total organic matter is widely used in bioprocess technology (especially for activated sludge wastewater treatment plants), for instance, in dry weight (DW) or of volatile suspended solids (VSS). However, a distinction between living biomass and dead organic material is not possible with this method. Thus, the proportion between the total cell count and this indirect biomass parameter can vary over a wide range. In the case of laboratory systems, a disadvantage of these simple methods is the comparably high amount of material.

CELLULAR COMPONENTS AS BIOMASS PARAMETERS

The use of cellular components as biomass parameters has to meet the requirements of (1) a ubiquitous presence in microbial cells, (2) a relatively constant proportion regardless of the species and the physiological state and (3) that it is characteristic of living cells.

Determination of DNA seems to be well suitable as such a parameter. DNA content of living bacterial cells is given as 3% of cellular dry weight (Brock et al., 1994) and is subject only to small fluctuations. A fluorometric assay using bisbenzimide (Hoechst 33258) (Brunk et al., 1979) has been applied with success to microorganisms in aquatic environments (Paul and Myers, 1982) and is recommended by Obst (1995) as a parameter of microbial biomass in environmental samples. This method is also useful for the estimation of biomass distribution in biofilm reactors.

As a first step in the determination of cellular components, these substances have to be extracted from the cells. In the present case, the extraction procedure was as follows: the biofilm material was removed from the silicone tubings as described above. After washing with phosphate buffer (0.05 M, pH 7.0), the biomass was disintegrated by sonication with an ultrasonic disintegrator (Sonopuls HD 60; Bandelin electronic, Berlin, Germany) with 100% power for 6 * 30 s (kept at 0°C by iced water). After centrifugation at 25,000 g, the supernatant was used for further examination.

The content of DNA in the supernatant was determined by means of bisbenzimide (Hoechst 33258) according to Obst (1995). The reaction mixture (3 ml) contained 50–200 μl of supernatant, citrate buffer (0.17 M, pH 7.0) and 10 μg ml^{-1} of bisbenzimide. Fluorescence was measured at an excitation wavelength of 345 nm and at an emission wavelength of 455 nm against a blank containing buffer and bisbenzimide by using a spectrofluorometer (RF 5001-PC; Shimadzu Europa GmbH, Duisburg, Germany).

The comparison of longitudinal distribution of DNA confirmed the heterogeneous microbial growth in the CFBR caused by gradients of substrate concentration (Figure 3). The distribution of fresh biomass of prokaryotes and eukaryotes, as calculated from the number and volume of bacterial and eukaryote cells, was in good agreement with the longitudinal gradient of the DNA content of biofilm extracts. This means that DNA is well suited as a marker of biomass for a simple and rapid determination of biomass distribution in biofilm reactors. Differences in spatial distribution of biomass (due to different modes

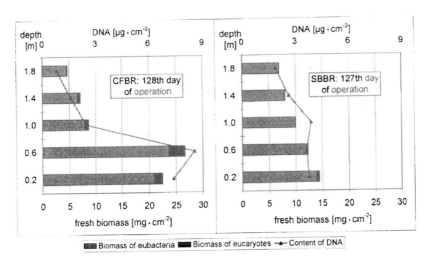

Figure 3 Comparison of longitudinal distribution of biomass of bacteria and eukaryotes and of DNA content in the two different operated biofilm reactors with addition of 4-chlorophenol (after 127/128 days of operation).

of operation or to different influent composition) became evident by DNA content to the same extent as by microscopic examination.

As expected, the biomass was more homogeneously distributed in the SBBR. However, because the SBBR was characterized by plug flow conditions, concentration gradients of substrates were formed by filling the reactor from the bottom to the top. This resulted in a weak gradient of biomass along the direction of flow also in the SBBR. As the reactor was drained against the direction of filling, sloughing due to the higher shear forces during this period and the displacement of the detached biomass to the lower part of the fixed bed (where it was retained), also affected the biomass distribution in the SBBR. Comparing the total biomass of both reactors, there was about 30% more biomass in the CFBR calculated both by means of DNA content and fresh biomass. The lower thickness of the biofilm in the SBBR was obviously caused by the higher shear stress during the filling and the draining periods.

Among the cellular polymers, the proteins represent the largest fraction with an average percentage of 50% of dry weight. However, as a parameter of biomass they are less suitable than DNA because of the high variability of protein content in microbial cells. Variations of the physiological state of microorganisms, for example starvation, may result in considerable changes of quantity and composition of proteins. In contrast to suspended (single-celled) cultures, biofilms are composed to a great extent of extracellular polymers (EPS) that include, besides polysaccharides, also a large amount of proteins (Christensen and Characklis, 1990; Jahn and Nielsen, 1995). Therefore, differences between viable cell count and protein content due to varying EPS composition and quantity are expected for different biofilms. However, the use of this parameter as a reference for enzyme activities is common in biochemistry and microbiology and is also proposed as a measure of specific activity of biofilms.

Mainly two photometric assays, the methods according to Lowry and Bradford, respectively, are well established in biochemistry and can be applied to the quantification of proteins in environmental samples (Lazarova and Manem, 1995). A risk is that the colorimetric reaction may interfere with numerous substances, for example humic substances. This can be overcome by modification of the standard assay (Frølund et al., 1995; Sperandio and Püchner, 1993). The application of Bradford's method to heterogeneous samples, such as biofilms or activated sludge, is limited by the comparatively high variability of color intensity due to different proteins. In this study, the proteins were detected in aliquots of the cell-free extracts by Lowry's method according to standard manuals (Süßmuth et al., 1987).

As can be seen from Figure 4, both the content of DNA and of protein samples from different depths of both reactors correlated well with the organic dry weight except the inflow section of CFBR (depth 0 m, C-0). An accumulation of dead organic matter from the inflowing sewage, especially humic

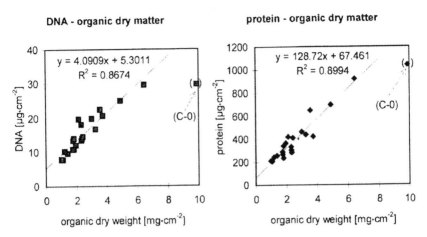

Figure 4 Correlation of DNA and protein content to organic dry weight of biofilm samples taken from several depths of the two different operated biofilm reactors. C-0 is the value of the sample taken from the inflow section of CFBR, depth 0 m (sampling after an operation period of 308 days with addition of 4-chlorophenol).

substances, might be responsible for the comparatively low DNA/organic DW or protein/organic DW ratio of the biofilm from this part. For all other samples, the percentage of protein ranged from 11 to 21% of organic dry weight, indicating a substantial contribution of dead organic material and/or carbohydrates (EPS) to the biofilm organic matter. Protein to DNA ratios of about 20:1 to 30:1 (for individual bacterial cells given with 10:1 to 25:1) make it probable that protein excretion by microorganisms into the biofilm matrix played an important role in the biofilm composition.

ESTIMATION OF BIOLOGICAL ACTIVITY

According to the substrate penetration, which decreases with increasing biofilm thickness due to internal mass transfer resistance, it is to be expected that the activity of biofilms does not proportionally increase with the biofilm thickness, i.e., biomass for example in terms of DNA. In the case of transport limitation, an increasing part of the biofilm remains inactive. Thus, the estimation and control of active biomass becomes very important for an optimization of biofilm reactor performance (Lazarova and Manem, 1995).

For characterization of the general metabolic activity of biofilms, some methods that are based on the measurement of the activity of different enzyme systems common among microorganisms are available. Because the organic substrates for heterotrophic bacteria to a great extent are macromolecular and not ready for incorporation into bacterial cells, hydrolytic enzymes are excreted by these microorganisms. In aquatic microbial ecology, the hydrolytic capacity of bacterial communities is examined by the application of bio-chemical methods to the determination of overall extracellular enzyme activity (Hoppe, 1993). The hydrolysis of fluorescein diacetate (FDA) by different ubiquitous extracellular esterases is used to a great extent as a marker of heterotrophic activity of aquatic microorganisms and in wastewater treatment (Nybroe et al., 1992; Lemmer et al., 1994).

For our investigation of biofilm activity we performed the FDA assay according to Obst (1995) by using homogenized (Ultra-Turrax; IKA-Labortechnik, Staufen, Germany) but not sonicated samples of biomass. The samples were diluted with phosphate buffer (0.06 M, pH 7.6) and, after addition of 100 μl of FDA (2 mg ml^{-1}), incubated for 1 hr at 20°C in a rotating mixer. The released fluorescein was measured photometrically at 490 nm in the supernatant after centrifugation at about 6000 g for 10 min. The enzymatic FDA hydrolysis was shown to correlate closely with the DNA content, suggesting similar proportions of active biomass in both reactors at different depths (Figure 5). This agrees with the close correlation of esterase activity to cell density of bacterial cultures (Schnürer and Rosswall, 1982) and to the ATP content (Stubberfield and Shaw, 1990).

We further compared the general heterotrophic activity with the occurrence and the activity of specialized microorganisms which are able to degrade (chloro-)-aromatics. Therefore, we measured the conversion of (chloro-)catechols to (chloro-)muconic acid by catechol-1,2-dioxygenase. This enzyme is involved in

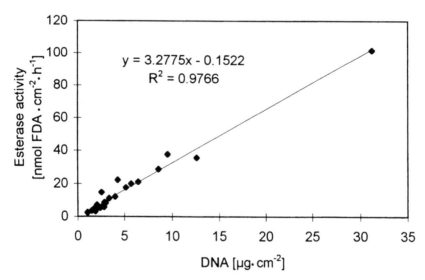

Figure 5 Correlation of enzymatic hydrolysis of fluorescein diacetate (FDA) as a marker of general heterotrophic activity and biomass in terms of DNA, for biofilm samples from different depths of CFBR and SBBR (after 168 days of operation with addition of 2,4,5-trichlorophenol).

the ortho-pathway of microbial catabolism of aromatic compounds. It has been suggested by some authors that the aerobic degradation of chlorophenols proceeds along this degradative sequence (Knackmuss and Hellwig, 1978; for a review, see also Häggblom, 1990; Commandeur and Parsons, 1994).

For the estimation of the activity of catechol-1,2-dioxygenase, an assay described by Reineke and Knackmuss (1984) was used. The reaction mixture contained (in 3 ml): 100 μmol Tris-HCl buffer (pH 8.0), 1 μmol of catechol or of the chloro-substituted analogue, 4 μmol EDTA, and 100–200 μl of cell-free extract (corresponding to a protein content from approximately 10–300 μg) according to the previously described procedure. Conversion of catechols to muconic acids was followed by the increase of absorbance at 260 nm. The molar absorption coefficients (ϵ) available from the literature (Dorn and Knackmuss, 1978) and the observed protein content were used for the calculation of the conversion rate in $nmol*h^{-1}*mg^{-1}$ protein.

Figure 6 shows the longitudinal distribution of general heterotrophic activity (hydrolysis of FDA) and of catechol conversion by catechol-1,2-dioxygenase in both reactors (the samples were taken after 168 days of operation with addition of 2,4,5-trichlorophenol). Although the esterase activity for the protein content was similar to some extent in both reactors and in the different reactor depths, the specific catechol-1,2-dioxygenase activity tended to increase with reactor depth. By this observation, an accumulation of microorganisms that were well adapted to chlorinated aromatics is made evident in the upper sections of the reactors. Contrary to the biomass distribution, the proportion of specialized microorganisms able to degrade (chloro-)aromatic compounds was heterogeneous along the fixed bed of both reactors, but it tended to increase with reactor depth both in the CFBR and in the SBBR.

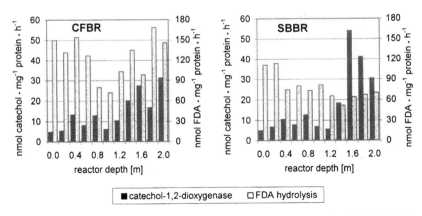

Figure 6 Longitudinal distribution of general heterotrophic activity (enzymatic FDA hydrolysis) and of conversion of catechol by catechol-1,2-dioxygenase in both reactors after 168 days of operation with addition of 2,4,5-trichlorophenol.

However, the ratio of catechol-1,2-dioxygenase activity to FDA hydrolysis was higher in samples taken from the SBBR (Figure 6). From this, the conclusion can be drawn that the proportion of specialized microorganisms or their activity was higher in the SBBR.

INVESTIGATION OF THE BACTERIAL COMMUNITY OF THE BIOFILM

According to the general acceptance of the larger rRNA molecules as phylogenetic markers, the rRNA approach became the most popular method to study biodiversity of complex microbial communities of natural and man-made environments (Pace et al., 1985; Olsen et al., 1986; Amann et al., 1990; Manz et al., 1994). Therefore, fluorescently labeled, rRNA-targeted oligonucleotide probes have been applied to examine the proportions between phylogenetic groups of Eubacteria in biofilms from different reactor depths by in situ hybridization. By the direct detection and identification of microorganisms in the biofilm samples it is possible to avoid the disadvantages of cultivation methods (low percentage of cultivable cells, selection pressure of media).

PREPARATION OF BIOFILM SAMPLES AND HYBRIDIZATION PROTOCOL

To quantify the proportions of different groups of Eubacteria in different sections of both reactors, the biofilms were removed from the tubings by sonication in a sonic water bath and subsequent scraping (as described in the previous section). After washing and homogenization, aliquots of suspended biofilm samples were fixed with freshly prepared paraformaldehyde solution and with absolute ethanol, respectively (according to Manz et al., 1994). The latter procedure was used to permeabilize cells of Gram-positive bacteria prior to hybridization. After fixation, the samples were washed and stored in PBS-ethanol (1:1) at $-20°C$.

For whole-cell hybridization of biofilm samples, the standard protocol described by Manz et al. (1994) was used. Prior to hybridization, the samples were homogenized by using an ultrasonic device (USD30; Emich Ultraschall GmbH, Berlin, Germany). To slow down the recoagulation of particles, one part of sodium pyrophosphate (2.8 g l^{-1}) was added to 10 parts of the homogenized sample. Between 1 and 20 µl of these biofilm samples was spread on glass slides (Paul Marienfeld, Bad Mergentheim, Germany) that previously had been coated with gelatin (0.1% gelatin, 0.01% chromium potassium sulfate) and dehydrated by sequential washes in 50%, 80%, and 98% ethanol (3 min each). The oligonucleotide probes (MWG Biotech, Ebersberg, Germany) used for examination of biofilm samples are listed in Table 2. For the probes

TABLE 2. Oligonucleotide Probe Data.

Probe	Specifity	Sequence 5'–3'	rRNA Target	Applied Stringency [Formamid Conc. (%)]	Reference
EUB338	Eubacteria	GCTGCCTCCCGTAGGAGT	16S	0	Amann et al., 1990
NON338	Negative control	CGACGGAGGGCATCCTCA	–	0	Amann et al., 1990
ALF1b	Alpha-subclass of Proteobacteria	CGTTCGYTCTGAGCCAG	16S	20	Manz et al., 1992
BET42a	Beta-subclass of Proteobacteria	GCCTTCCCACTTCGTTT	23S	35	Manz et al., 1992
GAM42a	Gamma subclass of Proteobacteria	GCCTTCCCACATCGTTT	23S	35	Manz et al., 1992
HGC69a	Gram-positive bacteria with high G + C-content	TATAGTTACCACCGCCGT	23S	20	Wagner et al., 1994
SRB385	Sulfate-reducing bacteria	CGGCGTCGCTGCGTCAGG	16S	35	Amann et al., 1992
CF319a + b	Cytophaga-Flavobacterium	TGGTCCGVTCTCAGTAC	16S	20	Wagner et al., 1994
NEU23a	Nitrosomonas sp.	CCCCTCGTGCACTCTA	16S	40	Wagner et al., 1995
GAM42a	Unlabeled competitor for BET42a	GCCTTCCCACATCGTTT	23S	35	Manz et al., 1992
BET42a	Unlabeled competitor for GAM42a	GCCTTCCCACTTCGTTT	23S	35	Manz et al., 1992
CTE	Unlabeled competitor for NEU23a	TTCCATCCCCCTCTGCCG	16S	40	Wagner et al., 1995

specific to the β- and the γ-subclasses of the Proteobacteria and to *Nitrosomonas* sp. (NEU), competitor probes have been used to improve the specifity of the hybridization, i.e., to block non-specific binding sites (Table 2). These probes had not been labeled with a fluorescent marker. The oligonucleotides have been diluted with twice-distilled, sterile water to a final concentration of 50 ng μl^{-1}.

To ensure optimal hybridization stringency, formamide was added to the hybridization buffer to the final concentrations given in Table 2. On each of the panels of glass slides, 8 μl of the hybridization buffer and 1 μl of the oligonucleotide probe with a concentration of 20–50 ng ml^{-1} were added. For all hybridizations, 50-ml polypropylene screw-top tubes served as hybridization (moisture) chambers. The slides were incubated for 1.5 hr at 46°C in a horizontal position. Subsequently, the hybridization mixture was rinsed with washing buffer that had a temperature of 48°C, and each slide was transferred into a vial with washing buffer of the same temperature. After 20 min, the slides were withdrawn, carefully rinsed with distilled water and air dried.

For the determination of the total cell number, the slides were stained with DAPI (Merck) at a final concentration of 1 $\mu g\ ml^{-1}$. The staining was performed subsequent to the in situ hybridization with 40 μl of DAPI on each panel for 15 min at room temperature in the dark. The biofilm samples were examined with a Jenalumar fluorescence microscope (Carl Zeiss, Jena). Combinations of specific filters were used for DAPI (exciter D 360/40, dichroic mirror 400 DC LP, emission filter D 460/50) and Cy3 (exciter HQ 535/50, dichroic mirror Q565 LP, emission filter HQ 610/75).

The slide was mounted with the embedding agent Citifluor AF1 (Citifluor Ltd., London, UK). To avoid the bleaching effects of Cy3 during DAPI examination, the hybridized cells were counted first and subsequently stained with DAPI. Per sample, about 1000–3000 cells were enumerated.

DATA ON BIOFILM COMMUNITIES OBTAINED BY IN SITU HYBRIDIZATION WITH OLIGONUCLEOTIDE PROBES

The following data refer to biofilm samples taken from the reactors at monthly intervals. The sampling points have been described above. In Figure 7, micrographs of a homogenized biofilm sample from the SBBR hybridized with oligonucleotide probes for the β-subclass of Proteobacteria and for the Cytophaga-Flavobacteria (CF) cluster are presented.

The examination of the composition of biofilm communities by in situ hybridization recovered 50–70% of the total cell count. The quantitative analysis revealed that the major proportion of the bacterial cells detected by means of the probes were members of the Proteobacteria (Figure 8). This comprises a very broad range of physiologically quite different taxa. The representatives of the β-subclass of Proteobacteria accounted for 10–50% of the hybridizable

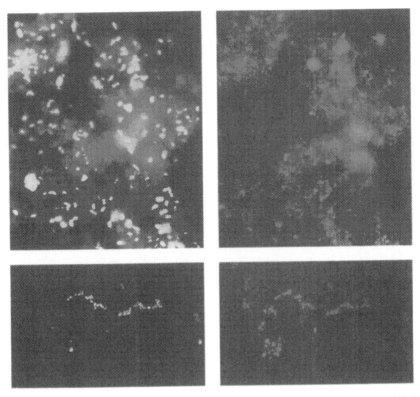

Figure 7 Hybridization of a biofilm sample from the SBBR with group-specific Cy3-labeled oligonucleotide probes. Left side, above: β-subclass of Proteobacteria; below: Cytophaga-Flavobacterium group. Right side: DAPI cell counts, ⊢———⊣ 10 μm.

cells, i.e., the number of bacteria detectable with the EUB probe and were the most abundant during the whole length of the experimental period. This finding corresponds to the results obtained from other wastewater treatment systems (Wagner et al., 1993; Manz et al., 1994).

The next important groups, according to cell abundance, were the Cytophaga-Flavobacteria (CF) cluster and the HGC group, respectively (Figure 8). Members of the CF group frequently have been found in many other wastewater treatment plants (Manz et al., 1994). Several members of the HGC group, such as *Arthrobacter* or *Rhodococcus,* are known for their capacity to decompose chlorinated phenols (Häggblom, 1990). Thus, their presence could be expected. Gram-positive bacteria with a high G + C content have been present in both reactors with proportions of 5–20% of the DAPI cell count in most of the samples. Particularly high proportions have been found in the uppermost segment of the CFBR. Between days 25 and 52 of MCP addition, the inflow concentration was increased from 20 to 60 mg l^{-1} MCP. In this pe-

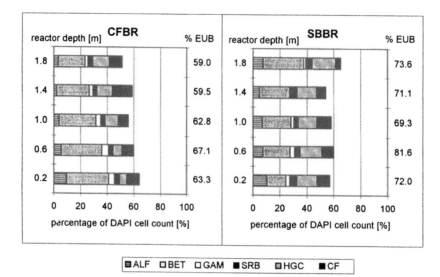

Figure 8 Composition of the microbial film at several sampling points along the reactor depth in the two different operated biofilm reactors with addition of 4-chlorophenol, as determined by in situ hybridization with fluorescently labeled, oligonucleotide probes specific for the α-, β- and γ-subclass of Proteobacteria (ALF, BET, GAM), for sulfate-reducing bacteria belonging to the Δ-subclass (SRB), for Gram-positive bacteria with high G + C content of DNA (HGC) and for the Cytophaga-Flavobacterium-cluster (CF) (sampling after 308 days of operation).

riod, the proportion of the HGC group sharply decreased. Thus, it is to be supposed that the members of the HGC group in the biofilm were more sensitive or less competitive at high MCP concentrations than other groups of the bacteria investigated.

The proportion of alpha-Proteobacteria in the total number of hybridizable cells (EUB) ranged between 2 and 15% in the CFBR. In the SBBR this proportion was higher on average (5–20%). In accordance with results of other researchers (Wagner et al., 1993; Manz et al., 1994), the γ-subclass of Proteobacteria was found to be insignificant in terms of cell numbers (usually below 5% of hybridizable cells).

Neither principal shifts in the proportions of principal groups of Eubacteria, nor conspicuous differences between the CFBR and the SBBR were observed over an experimental period of 9 months. The composition of microbial communities did not significantly differ from the population structure found in other wastewater treatment systems. Because the principal groups of Eubacteria comprise a very broad range of physiologically quite different taxa, changes in community structure relevant to the metabolic potential of the biofilms could not be detected by these probes. To what extent the composition of the microbial communities is different at the lower taxonomic levels still has to be explored with genus- and species-specific probes.

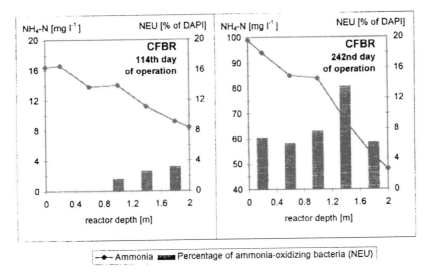

Figure 9 Comparison between the vertical profile of ammonia concentration and the longitudinal distribution of ammonia-oxidizing bacteria (as detected by NEU oligonucleotide probe) in the CFBR in an earlier (left) and a later (right) operational period.

For the nitrifying capacity, more detailed information can be obtained by the use of genus-specific probes, i.e., a probe specific to *Nitrosomonas* (NEU) (Wagner et al., 1995). The presence of nitrifiers was expected because of the low organic load and the high ammonia concentration in the inflow. In the first months of operation the proportion of *Nitrosomonas* was below 4% of total cell count, but in the following period it increased to 6–14% (corresponding to 10–20% of EUB). Correspondingly, a substantial part of the ammonia was oxidized. From Figure 9 it is evident that in the upper segment of the CFBR, where the NH₄ concentration rapidly decreased (i.e., a high nitrification rate was observed), the proportion of *Nitrosomonas* detected with the specific oligonucleotide probe was highest.

THE PROTOZOA AND METAZOA IN BIOFILMS

Biofilms are complex biotic systems in which food-web relationships are relevant. Although the biomass of the protozoa and metazoa may be low in comparison with the mass of bacteria, a biofilm reactor has to be considered as an ecosystem, the functional performance and stability of which may also depend on the eukaryotes. The interactions between bacteria and the other groups of organisms, thus, are of particular significance (see also Kinner and Curds, 1987; Lee and Welander, 1994).

EXAMINATION OF PROTOZOA AND METAZOA

Usually, the biomass was removed from a small piece of tubing with a length of 20–50 mm and suspended in a small amount of mineral water (free from carbon dioxide). The suspension was concentrated to a volume of 1–3 ml by centrifugation at low speed (3 min at 2000 rpm). An aliquot of this suspension (suspension A) was disintegrated with an Ultra-Turrax (30 s) at a very low speed. This more homogeneous material (suspension B) was used to enumerate the small protozoa in a Fuchs-Rosenthal counting chamber (depth 0.02 mm) at 400-fold magnification (microscope, Axioskop, Zeiss, Germany, with phase-contrast equipment). For counting larger organisms, 10 μl of suspension A was examined by using a Thoma chamber (depth 0.1 mm) with up to five replicates, as described by Augustin et al. (1989) for the investigation of protozoa and small metazoa in samples of activated sludge. A Kolkwitz chamber (volume 1 ml) was used to examine the large metazoa at 100-fold magnification.

Further calculations are based on the total number of individuals from all of the subsamples. Thus, no arithmetic means are presented. The numbers of individuals were related to the corresponding area of the tubings and were multiplied with the volumes of the different species (according to Foissner et al., 1995, or to average values obtained from microscopic measurement) to calculate the biomass (fresh mass).

UTILIZATION OF PROTOZOA AND METAZOA AS INDICATORS OF THE ECOLOGICAL CONDITIONS IN AND AROUND BIOFILMS

Protozoa and metazoa in biofilm reactors exert a direct influence on the composition and performance of the microbial community. Bacteria-grazing protozoa may decrease the number of bacteria that is relevant to the performance of the reactor (Lee and Welander, 1994), but their filtering activity also improves the quality of the effluent by the removal of suspended bacteria (Curds, 1982). In our experiments with addition of MCP to both reactors, a total number of 41 species of ciliates was found. Most of these occurred in the CFBR, whereas two thirds of the total number were observed in the SBBR. Altogether, 10 taxa of flagellates and rhizopods, and 11 taxa of metazoa, respectively, were found.

A succession in the composition of the eukaryotic community in the CFBR becomes evident from the aggregate biomass of both taxonomically different groups and physiological groups of ciliates (Figures 10 and 11). A more complex biotic structure of the biofilm involving several trophic levels can obviously develop in this type of reactor. The occurrence of carnivorous forms, such as rotifers of the genus *Cephalodella*, which reduce the number of bacterivorous protozoa, may have had a stabilizing effect on the total microbial biomass.

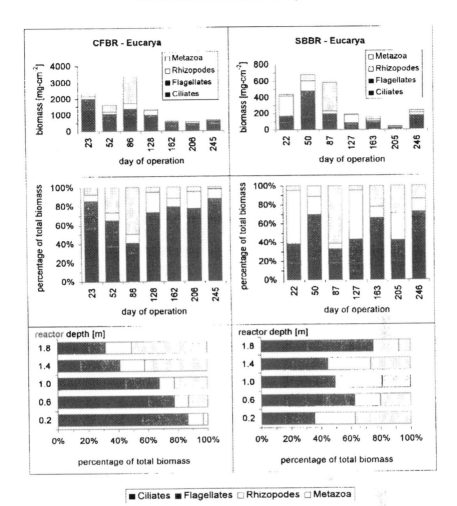

Figure 10 Top: Composition of eukaryote biomass in the CFBR and the SBBR at different sampling dates during the operation with addition of 4-chlorophenol (MCP). Middle: temporal changes in percentage composition of eukaryote biomass of both reactors during operation with MCP addition (averages of samples from 5 segments). Bottom: changes in percentage composition of eukaryote biomass across the longitudinal axis in both reactors (averages from 7 sampling dates).

Continuous-flow operation resulted also in a spatial succession toward a more complex community with several trophic levels. For example, the proportion of metazoa increased with the distance of flow, i.e., the height of the CFBR, which coincided with an increase in dissolved oxygen and a decrease in the MCP concentration (Figure 10). Among the protozoa, the ciliates normally were dominant in the CFBR. Flagellates have been found at high densities only in segments with reduced oxygen supply.

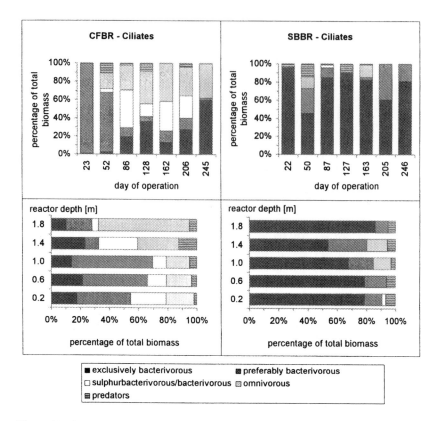

Figure 11 Above: temporal changes in percentage composition of ciliate biomass (for the type of feeding) in both reactors during operation with MCP addition (averages of samples from 5 segments). Below: percentage composition of the ciliate communities in different segments of the CFBR and the SBBR (averages from 7 sampling dates).

As for the composition of the community of eukaryotes, the SBBR was significantly different from the CFBR. There were temporal shifts in the dominance structure, with alternating maxima of different groups of protozoa (Figure 10). As an example, there were mass growths of amoebas followed by a maximum of the colorless flagellate Peranema.

The high abundance of heterotrophic flagellates and of amoebas has to be considered as a phenomenon characteristic of the starting phase in the operation of waste treatment plants or of overloading (Curds, 1982; Kinner and Curds, 1987). In the case of the SBBR, differences to the CFBR in the eukaryote community may be attributed both to the discontinuous feeding pattern (as described by Shih et al., 1996) and to the hydraulic stress during the filling and draining period. Similar to previously published observations

(Wobus et al., 1995), the eukaryote community structure in the SBBR seemed to be more susceptible to disturbances. However, bacterial growth may be promoted contrary to protozoan growth by these operating conditions (as supposed by Wilderer and Kaballo, 1995). Thereby, functional stability of such a biofilm system remains high. However, it has to be supposed that in a structurally unstable system the probability of a mass growth of invertebrates such as oligochaetes (of genus *Nais,* for example) or biofilm-feeding copepods such as *Paracyclops fimbriatus* is high. This might result in a critical reduction of active bacterial biomass.

For an assessment of the spatial changes in the composition of the eucaryote biomass, the ciliates were particularly well suited, considering their high quantitative proportions. The diversity of this group normally can be examined on the species level. Different species normally differ from one another in environmental requirements and, in particular, mode of feeding and type of food to such an extent that they represent good bioindicators.

In the CFBR, a spatial (longitudinal) succession of ciliates has been observed from the dominance of bacteria feeders toward a more complex community with several trophic levels (Figure 11). The control of bacterivorous protozoa by Suctoria predators such as *Podophrya fixa* and *Prodiscophrya collini* possibly increased the stability of the community in the CFBR. In the SBBR, predators and omnivorous ciliates did not play any significant role.

The dominance of species that exclusively or preferably feed on bacteria is obviously characteristic of treatment systems in the starting phase of operation. This confirms the hypothesis that no stable community could develop under the discontinuous flow conditions of the SBBR. As becomes evident from Figure 11, in which the longitudinal distribution of different ecological groups of ciliates is shown, the carnivorous and omnivorous forms were most abundant in the last, uppermost segment of the CFBR, which has the lowest organic load. In this segment, bacterivorous protozoa represented a comparatively low proportion of ciliates in accordance with the high grazing pressure exerted by these organisms.

The longitudinal distribution of the ciliates in the reactors also reflects corresponding differences in the oxygen supply. Species such as *Aspidisca cicada,* which have high demands on the oxygen concentration and thus indicate a high O_2 concentration, increased with the distance of water flow inside the CFBR. The O_2 concentration displayed the same longitudinal gradient. On the other hand, the occurrence of facultatively anaerobic ciliates such as *Spirostomum teres* and *Metopus es* in the first segment of the CFBR indicates that the oxygen supply here at least intermittently was insufficient. These two species mainly feed on bacteria, in particular sulfur bacteria, and have been found in the SBBR only very rarely.

In both reactors, sessile ciliates, mainly of the genera *Vorticella, Epistylis* and *Opercularia,* have been found. Different species groups were particularly abundant in the SBBR. In this context, the dominance of the α-mesosaprobic *Vorticella convallaria* indicates a good oxygen supply.

As the colonies of these peritrich ciliates rise above the surface of the biofilm, they probably profit from the higher velocity of current in the SBBR. In addition, the low thickness of the biofilm in this reactor increased the O_2-flux from the base, i.e., the membrane. The peritrich ciliates contribute substantially to the reduction in suspended solids and, in particular, bacteria. Because thick biofilms cannot easily be penetrated by substrate and oxygen and also tend to slough at very irregular intervals, a continuous removal of excess biofilm mass is desirable.

Considering the relevance of food-web relationships to the population dynamics of biofilm communities, the time resolution of examination should be high enough to identify oscillating predator/prey interactions. Biofilms "insensitive" to heavy grazing seem to be advantageous to maintain a high effluent quality of biofilm reactors. However, the conditions that are efficient in selecting these growth types are still unknown. On the other hand, it is to be supposed that a "moderate" concentration of eukaryote biomass facilitates mass transfer through the surface boundary layer of the biofilm due to increased turbulent diffusion by motion activities of the animals. Growth-stimulating effects likewise are attributed to the eukaryotes via the "forced rejuvenation" of the microbial film due to grazing and the release of growth-promoting dissolved substances.

CONCLUSIONS

Biofilms in fixed-bed reactors for the treatment of wastewaters containing xenobiotics may be examined according to different levels. The overall performance of the reactor is reflected by elimination rates and removal constants at a macroscopic level. Because these are significantly influenced by the longitudinal distribution of biomass, its activity and the population dynamics of the biofilm biocoenosis, these parameters were investigated.

As a marker of biomass, the DNA content seems to be well suited for the estimation of spatial distribution of microbial films. The distribution of biomass estimated by microscopic enumeration of prokaryotes and eukaryotes and measured for organic dry weight was in good agreement with the longitudinal gradients of DNA content. From all of these parameters, it became evident that the continuous operation of a plug flow fixed-bed reactor results in a stratified growth.

The functional heterogeneity of biofilms in the direction of flow may be elucidated by the estimation of biofilm activity. One approach is the comparison of removal rates (of organic load, expressed as DOC or COD, of special compounds, for example chlorophenols, or of ammonia, i.e., nitrification rates) related to a biomass parameter (DNA content). From these specific rates, the proportion of microorganisms that is active in the examined removal process can be estimated. However, this proportion is significantly influenced by the penetration of biofilms by dissolved substrates or by molecular oxygen due to the mass transfer resistance.

As another way of examination of biofilm activity, biochemical assays for the determination of the activity of enzymes common among microorganisms or of enzymes involved in a specific metabolic process can be conducted. By these means, it is possible to estimate the general or specific activity of biofilms *in vitro*. In this study, the catechol-1,2-dioxygenase activity and the general heterotrophic activity in terms of the enzymatic hydrolysis of FDA were compared in biofilm samples from different reactor depths. Thereby, different proportions of specialized microorganisms were made evident in different reactor depths.

According to the parameters mentioned above, the colonization of a plug flow fixed-bed reactor operated with continuous flow is characterized by a longitudinal heterogeneity. Biomass and microorganisms active in the removal of organic substances were enriched in the inflow section because of the formation of concentration gradients in the direction of flow. However, in the most distant section of this reactor, a thinner biofilm with a comparatively high proportion of specialized microorganisms (microorganisms with the capability to metabolize aromatic compounds and nitrifiers) was observed. Although the operation of an identical reactor in sequencing-batch mode resulted in a more homogeneous growth of biomass, a higher proportion of chloroaromatics-degrading microorganisms in the most distant part of the reactor became evident from the assay of catechol-1,2-dioxygenase activity.

Because the elimination of xenobiotics depends on their degradation by specialized bacteria or fungi, information about the biotic structure of microbial communities gains in significance for the control of reactor performance. Molecular biological techniques have rapidly become popular to study species composition or the proportions of different phylogenetic groups in microbial communities such as biofilms or activated sludge flocs (Wagner et al., 1993; Manz et al., 1994). From the present study, it is shown that the biotic structure of membrane-grown biofilms responsible for the elimination of chlorophenols did not significantly differ from the compostion of other microbial communities from wastewater treatment plants at the level of the principal groups of Eubacteria. For these groups, the membrane-grown biofilms of both reactors showed only a slight heterogeneity along the direction of flow. Groups such as Gram-positives with a high G + C content or as nitrifiers tended to preferably colonize the uppermost segment of the continuously operated reactor. However, a more detailed knowledge on the occurrence of distinct bacterial species or strains, and their metabolic activities is necessary to understand the organization of such complex biocoenoses. Consequently, there is a need of functional markers for specific activities and/or of probes specific for microorganisms with a distinct metabolic potential. This was the approach of Møller et al. (1996) for the examination of toluene-degrading pseudomonads in a multispecies biofilm, and of Schramm et al. (1996) for the investigation of nitrifying biofilms. An example for the examination of nitrifying bacteria in biofilms is also given in Chapter 5 of this book.

Although the biomass of the protozoa and metazoa was low in relation to the mass of bacteria, a biofilm reactor has to be considered as a complex ecosystem in which food-web relationships are also relevant. The interactions between bacteria and eukaryotic organisms are of particular significance to control the performance of a biofilm reactor. Grazing of the biofilm by mass growth of eukaryotes, such as oligochaetes (in our case of the genus *Nais*), snails (Palsdottir and Bishop, 1996) and protozoa (Lee and Welander, 1994), may result in a decrease of reactor performance.

Furthermore, the occurence of the protozoa and metazoa in biofilm reactors may be used as an indicator of the microenvironments in the biofilms (e.g., of the oxygen supply). Operating a biofilm reactor with continuous flow (CFBR), i.e., steady-state conditions, resulted in a spatial succession toward a more complex community with several trophic levels. Thereby, the stability of the biocoenoses in the CFBR increased. Because of the discontinuous mode of operation, the protozoan and metazoan community in the SBBR was significantly different from the CFBR and seemed to be more susceptible to disturbances.

Biofilm biomass is not only controlled by substrate concentration and shear stress but also by predator-prey interactions between eukaryotes and bacteria. On the other hand, the filtering activity of protozoa improves the quality of the effluent by the removal of suspended bacteria. As shown in this study, reactor design and the mode of operation significantly influence the quantity and activity of biomass, the biotic structure and their longitudinal distribution in a biofilm reactor. This emphasizes the advantage of an integrated, ecological approach to the examination of biofilms.

NOMENCLATURE

CF:	Cytophaga-Flavobacterium-cluster (or group-specific oligonucleotide probe)
CFBR:	Continuous-flow biofilm reactor
COD:	Chemical-oxygen demand
Cy3:	5,5'-Disulfo-1,1'-di(γ-carbophentynyl)-3,3,3',3'-tetramethylindolocarbocyanine-N-hydroxysuccinimide-ester
DAPI:	4',6-Diamidino-2-phenylindole
DOC:	Dissolved organic carbon
DW:	Dry weight

EPS:	Extracellular polymeric substances
EUB:	Eubacteria (or specific oligonucleotide probe)
FDA:	Fluorescein-diacetate
GC:	Gas chromatography
HGC:	Gram-positive bacteria with a high G + C content of DNA (or group specific oligonucleotide probe)
HPLC:	High-performance liquid chromatography
MCP:	4-(Mono-)chlorophenol
NEU:	Oligonucleotide probe specific for nitrifiying bacteria (*Nitrosomonas*)
PBS:	Phosphate-buffered saline
SBBR:	Sequencing batch biofilm reactor
SRB:	Sulfate-reducing bacteria (or group-specific oligonucleotide probe)
TCP:	2,4,5-Trichlorophenol

ACKNOWLEDGEMENTS

The present work was supported by the German national science foundation DFG—Deutsche Forschungsgemeinschaft (Grant Ro 1104/1-2). The authors acknowledge the examination of protozoa and metazoa by Mr. Sebastian Herborn. They are also grateful to Mrs. Inge Schachtschabel for her help in the experiments.

REFERENCES

Abrahamson, K. and T.M. Xie. 1983. Direct determination of trace amounts of chlorophenols in fresh water, waste water and sea water. *J. Chromatogr.* 279, 199–208.

Amann R.I., L. Krumholz and D.A. Stahl. 1990. Fluorescent-oligonucleotide probing of whole cells for determinative, phylogenetic, and environmental studies in microbiology. *J. Bacteriol.* 172, 762–770.

Amann, R.I., J. Stromley, R. Devereux, R. Key and D.A. Stahl. 1992. Molecular and microscopic identification of sulfate-reducing bacteria in multispecies biofilms. *Appl. Environ. Microbiol.* 58, 614–623.

Augustin, H., W. Foissner and R. Bauer. 1989. Die Zählung von Protozoen und kleinen Metazoen im Belebtschlamm. *Acta Hydrochim. Hydrobiol.* 17, 375–386.

Boller, M. and W. Gujer. 1986. Nitrification in tertiary trickling filters followed by deep-bed filters. *Water Res.* 20, 1363–1373.

Boller, M., W. Gujer and M. Tschui. 1994. Parameters affecting nitrifying biofilm reactors. *Water Sci. Technol.* 29(10/11), 1–11.

Bratbak, G. 1993. Microscope methods for measuring bacterial biovolume: epifluorescence microscopy, scanning electron microscopy, and transmission electron microscopy. In P.F. Kemp, B.F. Sherr, E.B. Sherr and J.J. Cole (eds.): *Handbook of methods in aquatic microbial ecology.* Lewis Publishers, Boca Raton, 309–317.

Brock, T.D., M.T. Madigan, J.M. Martinko and J. Parker. 1994. *Biology of microorganisms.* Prentice-Hall Inc., Englewood Cliffs, New Jersey.

Brunk, C.F., K.C. Jones and T.W. James. 1979. Assay of nanogram quantities of DNA in cellular homogenates. *Anal. Biochem.* 92, 497–500.

Bryers, J.D. and W.G. Characklis. 1990. Biofilms in water and wastewater treatment. In W.G. Characklis and K.C. Marshall (eds.): *Biofilms.* John Wiley, New York, 671–696.

Christensen, B.E. and W.G. Characklis. 1990. Physical and chemical properties of biofilms. In W.G. Characklis and K.C. Marshall (eds.): *Biofilms.* John Wiley, New York, 93–130.

Commandeur, L.C.M. and J.R. Parsons. 1994. Biodegradation of halogenated aromatic compounds. In: C. Ratledge (ed.): *Biochemistry of microbial degradation.* Kluwer Academic Publ., Dordrecht, 423–458.

Curds, C.R. 1982. The ecology and role of protozoa in aerobic sewage treatment processes. *Ann. Rev. Microbiol.* 36, 27–46.

Dorn, E. and H.-J. Knackmuss. 1978. Chemical structure and biodegradability of halogenated aromatic compounds: substituents effects on 1,2-dioxygenation of catechol. *Biochem. J.* 174, 85–94.

Fletcher, M. 1992. The measurement of bacterial attachment to surfaces in static systems. In: L.F. Melo, T.R. Bott, M. Fletcher and B. Capdeville (eds.): *Biofilms— science and technology.* Kluwer Academic Publ., Dordrecht, 603–614.

Foissner, W., H. Berger, H. Blatterer and F. Kohmann. 1995. *Taxonomische und ökologische Revision der Ciliaten des Saprobiensystems. Band IV: Gymnostomatea,* Loxodes, *Suctoria.* Inform. Ber. Bayer. Landessamt Wasserwirtschaft, München.

Frølund, B., R. Palmgren, K. Keiding and P.H. Nielsen. 1995. Extraction of extracellular polymers from activated sludge using a cation exchange resin. *Water Res.* 30, 1749–1758.

Häggblom, M.M. 1990. Mechanisms of bacterial degradation and transformation of chlorinated monoaromatic compounds. *J. Basic Microbiol.* 30, 115–141.

Harremoës, P. 1978. Biofilm kinetics. In: R. Mitchell (ed.): *Water pollution microbiology, Vol. 2.* John Wiley, New York, 71–111.

Hoppe, H.-G. 1993. Use of fluorogenic model substances for extracellular enzyme activity (EEA) measurement of bacteria. In: P.F. Kemp, B.F. Sherr, E.B. Sherr and J.J. Cole (eds.): *Handbook of methods in aquatic microbial ecology.* Lewis Publishers, Boca Raton, 423–431.

Irvine, R.L., T.P. Fox and R.O. Richter. 1977. Investigation of fill and batch periods of sequencing batch biological reactors. *Water Res.* 11, 713–717.

Jahn, A. and P.H. Nielsen. 1995. Extraction of extracellular polymeric substances (EPS) from biofilms using a cation exchange resin. *Water Sci. Technol.* 32(8), 157–164.

Kaballo, H.-P., Y. Zhao and P.A. Wilderer. 1995. Elimination of *p*-chlorophenol in biofilm reactors—a comparative study of continuous flow and sequenced batch operation. *Water Sci. Technol.* 31(1), 51–60.

Kinner, N.E. and C.R. Curds. 1987. Development of protozoan and metazoan communities in rotating biological contactor biofilms. *Water Res.* 21, 481–490.

Knackmuss, H.-J. and M. Hellwig. 1978. Utilization and cooxidation of chlorinated phenols by *Pseudomonas* sp. B13. *Arch. Microbiol.* 117, 1–7.

Ladd, T.I. and J.W. Costerton. 1990. Methods for studying biofilm bacteria. In: R. Grigorova and J.R. Norris (eds.): *Methods in microbiology, Vol. 22, Techniques in microbial ecology.* Academic Press, London, 285–307.

Lazarova, V. and J. Manem. 1995. Biofilm characterization and activity analysis in water and wastewater treatment. *Water Res.* 29, 2227–2245.

Lee, N.M. and T. Welander. 1994. Influence of predators on nitrification in aerobic biofilm processes. *Water Sci. Technol.* 29(7), 355–363.

Lemmer, H., D. Roth and M. Schade. 1994. Population density and enzyme activities of heterotrophic bacteria in sewer biofilms and activated sludge. *Water Res.* 28, 1341–1346.

Manz, W., M. Wagner, R. Amann and K.-H. Schleifer. 1994. In situ characterization of the microbial consortia active in two wastewater treatment plants. *Water Res.* 28, 1715–1723.

Manz, W., R. Amann, W. Ludwig, M. Wagner and K.-H. Schleifer. 1992. Phylogenetic oligodeoxynucleotide probes for the major subclasses of proteobacteria: problems and solutions. *Sys. Appl. Microbiol.* 15, 593–600.

Møller, S., A.R. Pedersen, L.K. Poulsen, E. Arvin and S. Molin. 1996. Activity and three-dimensional distribution of toluene-degrading *Pseudomonas putida* in a multispecies biofilm assessed by quantitative in situ hybridization and scanning confocal laser microscopy. *Appl. Environ. Microbiol.* 62, 4632–4640.

Nybroe, O., P.E. Jørgensen and M. Henze. 1992. Enzyme activity in waste water and activated sludge. *Water Res.* 26, 579–584.

Obst, U. 1995. *Enzymatische Tests für die Wasseranalytik.* R. Oldenbourg, München, Wien.

Olsen, G.J., D.J. Lane, S.J. Giovannoni, N.R. Pace and D.A. Stahl. 1986. Microbial ecology and evolution: a ribosomal RNA approach. *Ann. Rev. Microbiol.* 40, 337–365.

Pace, N.R., D.A. Stahl, D.J. Lane and G.J. Olsen. 1985. The analysis of natural microbial populations by ribosomal RNA sequences. *Adv. Microbiol. Ecol.* 9, 1–55.

Palsdottir, G. and P.J. Bishop. 1996. Nitrifying biotower upsets due to snails and their control. In: *Proceedings of Third International IAWQ Special Conference on Biofilm Systems,* 1996, Copenhagen.

Paul, G.H. and B. Myers. 1982. Fluorometric determination of DNA in aquatic microorganisms by use of Hoechst 33258. *Appl. Environ. Microbiol.* 43, 1393–1399.

Porter, K.G. and Y.S. Feig. 1980. The use of DAPI for identifying and counting aquatic microflora. *Limnol. Oceanogr.* 25, 943–948.

Reineke, W. and H.-J. Knackmuss. 1984. Microbial metabolism of haloaromatics: isolation and properties of a chlorobenzene-degrading bacterium. *Appl. Environ. Microbiol.* 47, 395–402.

Schnürer, J. and T. Rosswall. 1982. Fluorescein diacetate hydrolysis as a measure of total microbial activity in soil and litter. *Appl. Environ. Microbiol.* 43, 1256–1261.

Schramm, A., L.H. Larsen, N.P. Revsbech, N.B. Ramsing, R. Amann and K.-H. Schleifer. 1996. Structure and function of a nitrifying biofilm as determined by in situ hybridization and the use of microelectrodes. *Appl. Environ. Microbiol.* 62, 4641–4647.

Shih, C.-C., M.E. Davey, J. Zhou, J.M. Tiedje and C.S. Criddle. 1996. Effects of phenol feeding pattern on microbial community structure and cometabolism of trichloroethylene. *Appl. Environ. Microbiol.* 62, 2953–2960.

Siegrist, H. and W. Gujer. 1985. Mass transfer mechanisms in a heterotrophic biofilm. *Water Res.* 19, 1369–1378.

Sperandio, A. and P. Püchner. 1993. Bestimmung der Gesamtproteine als Biomasse-Parameter in wäßrigen Kulturen und auf Trägermaterialien aus Bio-Reaktoren. Modifizierte Methode nach Lowry—Eine praktikable Methode in der Umweltanalytik. *gwf-Wasser-Abwasser* 134, 482–485.

Stubberfield, L.C.F. and P.J.A. Shaw. 1990. A comparison of tetrazolium reduction and FDA-hydrolysis with other measures of microbial activity. *J. Microbiol. Methods* 12, 151–162.

Süßmuth, R., J. Eberspächer, R. Haag and W. Springer. 1987. *Biochemisch-mikro-biologisches Praktikum*. Georg Thieme Verlag Stuttgart, New York.

Wagner, M., R. Amann, H. Lemmer and K.-H. Schleifer. 1993. Probing activated sludge with oligonucleotides specific for proteobacteria: inadequacy of culture-dependent methods for describing microbial community structure. *Appl. Environ. Microbiol.* 59, 1520–1525.

Wagner, M., G. Rath, R. Amann, H.-P. Koops and K.-H. Schleifer. 1995. In situ iden-tification of ammonia-oxidizing bacteria. *Sys. Appl. Microbiol.* 18, 251–264.

Wagner, M., R. Erhart, W. Manz, R. Amann, H. Lemmer, D. Wedi and K.-H. Schleifer. 1994. Development of an rRNA-targeted oligonucleoide probe specific for the genus *Acinetobacter* and its application for in situ monitoring in activated sludge. *Appl. Environ. Microbiol.* 60, 792–800.

Wilderer, P.A. 1992. Sequencing batch biofilm reactor technology. In: M.R. Ladisch and A. Bose (eds.): *Harnessing biotechnology for the 21st century*. American Chemical Society, NY, 475–479.

Wilderer, P.A. and H.-P. Kaballo. 1995. Influence of nonsteady-state process condi-tions on biofilm dynamics—concept, findings and research needs. In: *Proceed-ings of International IAWQ Conference Workshop: Biofilm structure, growth and dynamics*, 1995, Noordwijkerhout, The Netherlands, 92–101.

Wilderer, P.A., I. Röske, A. Ueberschär and L. Davids. 1993. Continuous flow and se-quenced batch operation of biofilm reactors: a comparative study of shock load-ing responses. *Biofouling* 6, 295–304.

Wobus, A. and I. Röske. 1999. Reactors with membrane-grown biofilms: their capac-ity to cope with fluctuating inflow conditions and with shock loads of xenobi-otics. *Water Res.*, 34(1), 279–287, 2000.

Wobus, A., S. Ulrich and I. Röske. 1995. Degradation of chlorophenols by biofilms on semi-permeable membranes in two types of fixed bed reactors. *Water Sci. Tech-nol.* 32(8), 205–212.

CHAPTER 12

Quantitative Microscopy in Biofilm Studies

MARTIN KUEHN
ULRICH SCHINDLER
PETER A. WILDERER
STEFAN WUERTZ

INTRODUCTION

BIOFILMS are ubiquitous and occur wherever sufficient water, inorganic and organic substrates can be found. They form at interfaces and represent the sessile form of life for microorganisms compared with the planktonic stage. Despite the growing recognition that biofilms represent the predominant form of life for microorganisms, few attempts have been made to investigate biofilms noninvasively and quantitatively. With the advent of confocal laser scanning microscopy (CLSM) and its introduction to medical and finally environmental sciences, it has become possible to look at biological structures without having to prepare samples for analysis. This has been an improvement over electron microscopy, which has been associated with introducing artifacts. Until now, most microscopic studies of biofilms have been descriptive without attempting to derive quantitative information from the acquired images as reviewed by Caldwell et al. (1992a, 1992b). In some cases a small number of selected images were subjected to data analysis (Lawrence et al., 1989; Lawrence et al., 1991; Caldwell et al., 1992a, 1992b; Bloem et al., 1992). Here we present a method for routine measurements of biofilms using automated image acquisition and image analysis (Kuehn et al., 1998).

MATERIALS AND METHODS

MICROSCOPY

Images were generated with the commercial Zeiss CLSM 410 system by scanning with a He-Ne laser at 543 nm or an Ar laser at 488 nm and 514 nm.

The motorized scanning stage was controlled by the Zeiss software. This can be performed automatically by a user-specified macro procedure. Images were saved on an external hard drive disk with 1-GByte storage area.

IMAGE PROCESSING

The digital image processing was performed with a Leica Quantimet 570 computer. By using image analysis, geometric measurements can be performed on stored images originating from various sources. This allows for complex tasks, such as calculating the area or volume taken up by biological structures, counting particles, the analysis of particle shapes and sizes, or the measurement of distances between particles, to be performed.

The main functions of any image processing are:

- image acquisition
- gray image analysis
- detection and binary image analysis
- measurements

During image acquisition one adjusts gain (contrast) and offset (intensity) levels, the degree of illumination, and image integration to suppress background noise. During gray image processing the image can be expanded, edges can be redefined and boundaries set. Morphological transformations are achieved by erosion and gradient determination and by arithmetic operations like adding and subtracting of individual images.

Detection involves setting of thresholds that are either freely adjustable or automatic. Images are divided into white, gray and black areas and improved by opening and closing, by identifying the ends and crossover points of lines, and by skeletonizing, drawing and cutting. Then, images are scored as 0 or 1 (black or white) for further processing.

Measurements involve procedures such as calibration, setting of image and measurement frames (area of interest), assignment of area and particle specific measurement parameters, and statistics with gray value histograms and profiles.

Image analysis is performed with command line scripts unlike compiling computer languages such as FORTRAN or PASCAL. This allows a stepwise interactive protocol. The syntax of interpreter based language is similar to that of popular PC languages like QBASIC. Here the user languages specified by the Leica Quantimet 570 computer were QUIN (Quantimet under Windows) and QUIPS (Quantimet Image Processing System). QWIN is the interpreter language and facilitates interactive procedures. QUIPS is based on QWIN and enables the user to program macros.

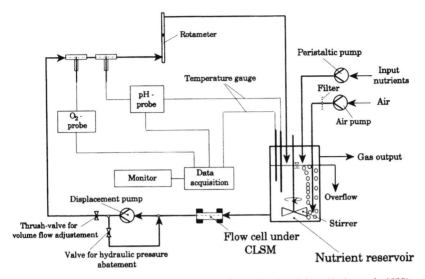

Figure 1 Schematic view of fermentor setup and flow cell (adapted from Kuehn et al., 1998).

FLOW CELL SYSTEM

Bacteria were grown in an aerated and stirred fermentor. The fermentor was connected to a flow cell by tubing (Figure 1). A pressure-independent pump circulated the fluid from the fermentor through the cell and back into the fermentor. A constant flow free of pulsations was maintained by a hydraulic feedback loop encompassing the displacement pump. The cell (Figure 2) consisted of two coverslips with a thickness of 0.2 mm each that were glued with silicon to a stainless steel frame 46 mm long, 8 mm wide, and 2.7 mm high. The area of the flow cell cross section measured 21.6 mm^2. The wetted perimeter of the cell was 21.4 mm. The complete system was sterilized in an autoclave, except the displacement pump and the pH and temperature probes, which were cleaned with 0.1$\times$ acetyl hydroperoxide followed by rinsing with autoclaved distilled water before use.

BIOFILM GROWTH

An overnight culture of *P. fluorescens* was inoculated into 0.1$\times$ standard Luria broth. The bacterial suspension that developed in the fermentor was circulated at a constant flow rate of $Q_C = 3.0$ l/hr through the system. After 24 hr the biofilm on the coverslips was subjected to CLSM analysis. Cells could be visualized on the basis of their autofluorescence excited by a laser beam at the wavelength of 488 nm.

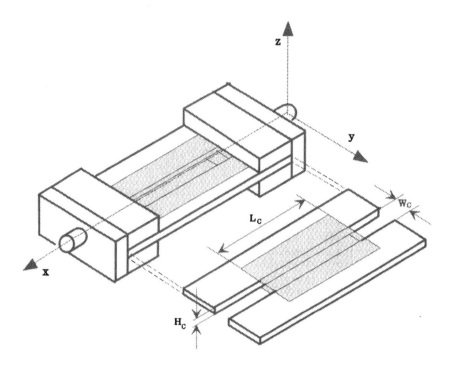

Channel length	$L_C = 46$ mm
Channel width	$W_C = 8$ mm
Channel height	$H_C = 2,7$ mm
Channel : Area of cross section	$A_C = 21,6$ mm^2
Channel : wetted perimeter	$P_C = 21,4$ mm

Figure 2 Detailed view of flow cell (after Kuehn et al., 1998. Reprinted with permission from American Society for Microbiology, © 1998).

RESULTS

AXIAL ABERRATION

Optical sectioning of biofilms is possible in the horizontal (xy) and sagittal (xz) direction as outlined in Figure 3. In the horizontal direction spherical deviations are negligible. Sagittal sections, however, result in serious axial aberrations as illustrated in Figure 4. Fluorescent beads of an average diameter of 2 μm were excited at 488 nm and visualized by using a 100×/1.3 oil im-

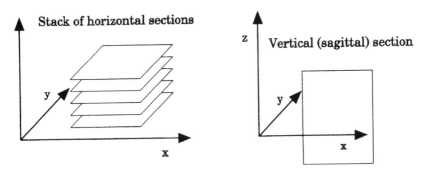

Figure 3 Schematic view of horizontal and sagittal sections taken with a confocal laser scanning microscope CLSM.

mersion objective lens. From the examples given in Figure 4, the relative error based on the diameter of a fluorescent latex bead can be calculated as

$$r = \frac{3.5 - 2.0}{2.0} = 0.75 \equiv 75\%$$

Because the software package that runs the CLSM does not contain a corrective algorithm for axial aberration, the objective has to be calibrated by the user applying fluorescent beads and manual correction utilizing the $z{:}xy$ ratio command in the z-scan mode of the CLSM interpreter language.

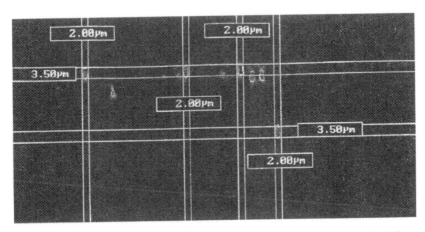

Figure 4 Axial aberration during sagittal (xz-) scanning of fluorescent microbeads (bead diameter: 2 μm).

BIOFILM DEVELOPMENT

The attachment of cells was followed microscopically on the basis of the fluorescence of *P. fluorescens*. Taking into account the flow rate in the flow channel Q_C, the mean bulk fluid velocity in the channel can be calculated as

$$v_m = Q_C/A_c \tag{1}$$

The Reynolds number, Re, is dimensionless and is characterized by the ratio of inert forces to frictional forces. It is used to describe the type of flow conditions in the channel.

$$Re = 4r_h v_m/v \tag{2}$$

with

$$r_h = A_c/P_c \tag{3}$$

where:

Q_C = flow rate in the flow channel
A_C = cross section area of channel
r_h = hydraulic radius
P_C = wetted perimeter of the flow channel
v = coefficient of kinematic viscosity of the fluid (v_{water} = 0.01 cm^2/s at 20°C)

Under the experimental conditions used v_m was 3.5 cm/s associated with a Reynolds number Re $\approx$ 140; therefore, the adhesion and growth of bacteria took place under laminar flow conditions and was influenced by frictional forces too. After 13 hr cells had already settled along the channel walls. It is known that cell density and biofilm development are influenced by the flow near an interface. Images were acquired in the channel region to investigate this phenomenon (Figure 5). The domain of interest for measurements had the following geometry: length in direction x is 639 μm, length in direction y is 3067 μm, and length in direction z is 20 μm.

DATA ACQUISITION

Confocal images do not suffer from stray light originating from other focal planes due to the existence of a pinhole. Images were acquired in stacks as xy sections with a vertical step interval $\Delta z = 1$ μm using a 100×/1.3 oil immersion lens (Figure 6). About 1900 images consisting of 512 × 512 pix-

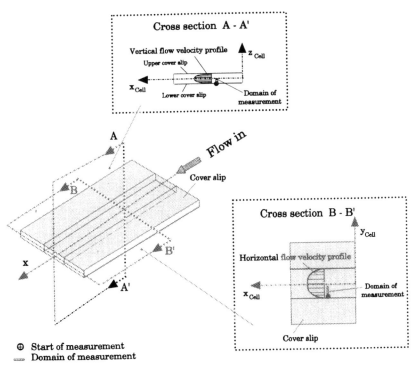

Figure 5 Schematic view of the flow cell indicating fluid velocity profiles and domain of measurement (after Kuehn et al., 1998. Reprinted with permission from American Society for Microbiology, © 1998).

els were captured. Based on a calibration to the lens, this corresponded to a reference length lr of 127.8 μm. Each image package consisted of $le = 10$ images, and each image stack contained $ke = 2$ packages. The desired number of images in the x and y direction determines the total number ne of stacks. By using the variables dx and dy, the distances between image stacks can be chosen independently (Figure 6). Based on a user-specified macro procedure, image acquisition could be automated (Figure 7).

NUMERICAL TREATMENT

Optical sections featuring *P. fluorescens* cell accumulation on the surface area (substratum) in its early phase of development were affected by light reflection. Consequently, the images acquired at position $z = 0$ μm could not be used for image analysis (Figure 8). Therefore, a numerical approximation procedure had to be integrated into the image analysis software, allowing a reasonable curve fitting directly on the substratum (Kuehn et al., 1998). For

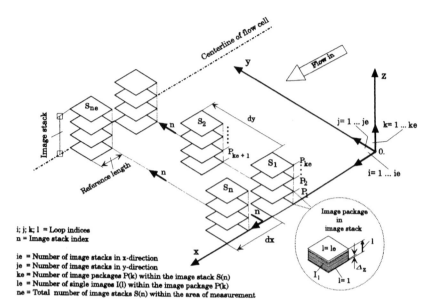

i; j; k; l = Loop indices
n = Image stack index

ie = Number of image stacks in x-direction
je = Number of image stacks in y-direction
ke = Number of image packages P(k) within the image stack S(n)
le = Number of single images I(l) within the image package P(k)
ne = Total number of image stacks S(n) within the area of measurement

Figure 6 Schematic view of automated image acquisition using the Zeiss CLSM 410 microscope (after Kuehn et al., 1998. Reprinted with permission from American Society for Microbiology, © 1998).

the regimen of the curve immediately at the surface area (Figure 9), an approach can be formulated by a polynomial of degree 6 that fits well with the measured datum points.

$$p(z) = \sum_{i=0}^{6} a_i \cdot z^i \qquad (4)$$

where a_i are the coefficients of the polynomial.

The images can be analyzed numerically by integration to define an approximate accumulative biovolume in a defined biofilm volume. The procedure is based on the function $f(x,y,z)$ in a given volume V.

$$\iiint_V f(x,y,z)dV = \int_z F(z)dz \qquad (5)$$

The functional relationship $F(z) = \iint_{x,y} f(x,y,z)dydx$ that describes bacterial accumulation on a surface area x, y as a function of z is given as a table by the image analysis program Quantimet without describing the basic function $F(z)$ mathematically. For this reason the integral cannot be resolved ana-

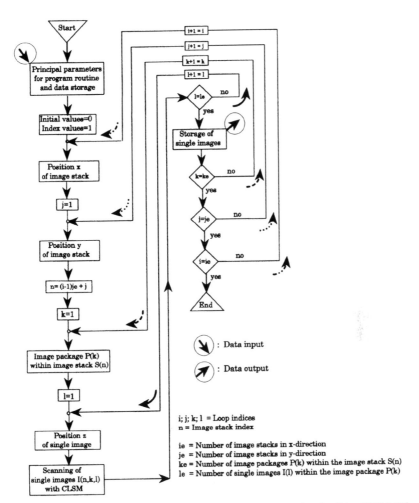

Figure 7 Flow chart of the macro routine for image acquisition using the Zeiss CLSM 410 computer.

lytically. However, by using the numerical approximation procedure shown in Equation (6) the integrand could be found.

$$\int_{za}^{ze} F(z)dz \approx \sum_{m=1}^{me} c_m \cdot F(z_m) \tag{6}$$

where c_m = nonnegative weight coefficients, and $me = ke \cdot le \equiv$ number of measured points (images) used for the numerical integration (see Figures 6 and 7).

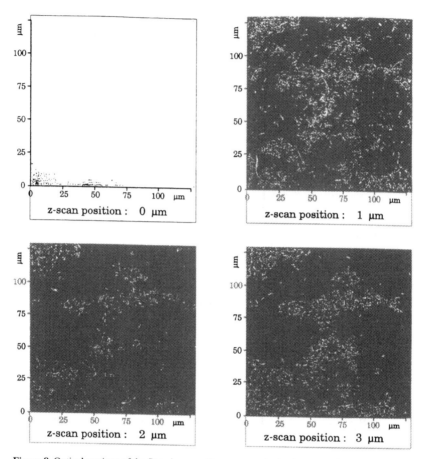

Figure 8 Optical sections of the *Pseudomonas fluorescens* colonization at varying channel depths using the CLSM. At a position close to the substratum images are affected by light reflection.

This involves the integral of a function in a finite interval between za and ze with the weighted sum of a number of measured values per image stack $F(z_1), F(z_2), \cdots, F(z_{me})$ for the sectional depths of $z_m = \in [za, ze]$.

Here, we used a pragmatic approach to derive a formula for numerical integration by following the trapezoidal rule, $\int_{za}^{ze} F(z) dz$, is considered the area under the curve $F(z)$ within the numerical boundaries za and ze. This method has the advantage of being independent of the number of supporting points and is valid for well-conditioned $F(z)$. As could be seen from measurements the datum points of the colonization curves, $F(z)$ do not show any divergence and indicate a continuous trend in the diagrams.

To illustrate the approach a test function is shown in Figure 10. The graph is similar to the observed bacterial adhesion in Figure 9. The analytical solu-

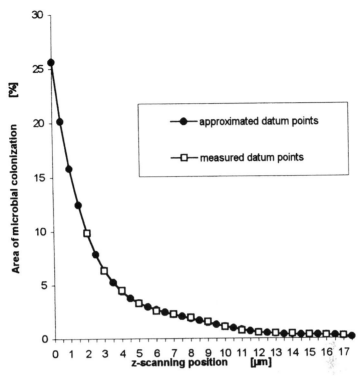

Figure 9 Typical colonization curve showing measured bacterial growth and extrapolated values. The numerical approximation procedure was a least-square fit of measured data points to a polynomial function of degree 6 (after Kuehn et al., 1998. Reprinted with permission from American Society for Microbiology, © 1998).

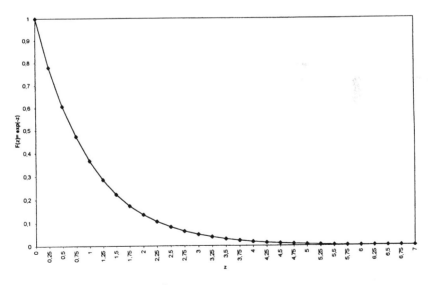

Figure 10 Test function $F(z) = e^{-z}$.

tion is known and can be used to assess the success of numerical approximation. The intervals $\Delta z = z_{m+1} - z_m$ between individual datum points z_m do not have to be equidistant when using the trapezoidal rule. Therefore, the biovolume V_n enclosed in each stack S_n at a certain xy position can be calculated by means of the numerical integrand by using the matrix equation:

$$
V_n = I_{num} \approx [(z_2 - z_1)(z_3 - z_2) \cdots (z_{me} - z_{me-1})] \begin{bmatrix} F(z_1) + F(z_2) \\ F(z_2) + F(z_3) \\ \cdot \\ \cdot \\ \cdot \\ F(z_{me-1}) + F(z_{me}) \end{bmatrix} /2 \quad (7)
$$

For $\Delta z = (z_2 - z_1) = (z_3 - z_2) = \cdots = (z_{me} - z_{me-1}) = $ constant, Equation (7) can be simplified as follows:

$$
V_n = I_{num} \approx \Delta z \begin{bmatrix} F(z_1) + F(z_2) \\ F(z_2) + F(z_3) \\ \cdot \\ \cdot \\ \cdot \\ F(z_{me-1}) + F(z_{me}) \end{bmatrix} /2 \quad (8)
$$

By using Equation (9), all stack volumes V_n calculated within the image stacks S_n and summarized over the area of interest (Figures 5 and 6) allow us to assess the total biovolume V_t accumulated in the domain of measurement.

$$
V_t = \frac{1}{V_r} \sum_{n=1}^{ne} V_n \quad (9)
$$

where V_r = reference volume $\equiv l^3$, and l_r = reference length; it depends on the magnification of the lens in use.

ERROR CAUSED BY THE TRAPEZOIDAL SUM

The error following the trapezoidal rule for integration is proportional to Δz^2, i.e., if one halves the partial interval length Δz, the absolute error of the integral ($I_{num} - I_{analyt}$) is reduced by a factor of 4.

The analytical integration of a test function shown in Figure 10 gives the following exact result:

$$
I_{analyt} = \int_0^7 e^{-z} dz = -e^{-z} \Big|_0^7 = -9.11881 \times 10^{-4} + 1.0 = 0.999088 \quad (10)
$$

Applying Equation 8 the numerical integrand of the function depicted in Figure 10 can be calculated:

$$I_{num} \approx 1.004286$$

with $za = 0$; $ze = 7$; $\Delta z = 0.25$ and the number of data points $m_e = 16$.
A comparison of both results yields the relative error:

$$r = \frac{I_{num} - I_{analyt}}{I_{analyt}} = \frac{1.004286 - 0.999088}{0.999088} = 0.0052 \equiv 0.5\% \quad (11)$$

Finally, the principal steps of the macro routine integrated into the image analysis software are shown in the flowchart of Figure 11. The actual program is available from the authors on request.

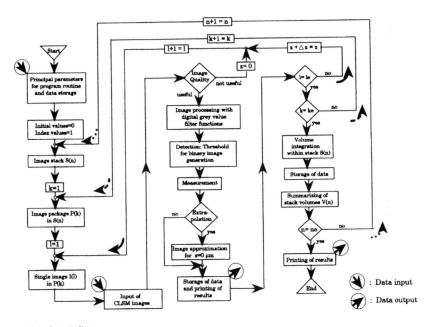

k; l = Loop indices
n   = Image stack index

ke = Number of image packages P(k) within the image stack S(n)
le = Number of single images I(l) within the image package P(k)
ne = Total number of image stacks S(n) within the area of measurement

Figure 11 Flowchart of the macro routine for image analysis using the Leica Quantimet 570 computer (after Kuehn et al., 1998. Reprinted with permission from American Society for Microbiology, © 1998).

DISCUSSION

Horizontal sections of a biofilm taken by a confocal laser-scanning microscope can be stored in a digital format and used to calculate the area taken up by biological structures. Here we used the autofluorescence of a strain of *Pseudomonas fluorescens* to follow the development of a biofilm in a flow cell. By integrating a number of optical sections, one can obtain an estimate of the biovolume in a biofilm. We chose a geometrical approach to be independent of the number of data points. Instead of using autofluorescence, one could stain the biofilm and count specific cells or label other biological structures (Kuehn et al., 1998). Other approaches have relied on the delineation and edge detection of single cells after staining (Moller et al., 1995). Our method does not depend on the detection of individual cells, which is difficult in biofilms containing several cell layers. By computing the area or volume taken up by specific cells based on their fluorescence and comparing it to the overall cellular volume it will be possible to construct depth-resolved profiles of any microbial population without the need for cell counts. This should simplify the study of biofilms considerably because automated counts can be performed randomly.

It should be realized, however, that the method does not represent a true three-dimensional (3D) approach. Currently, the costs of 3D imaging software are considerable, and there is no ready-to-use program for biofilms on the market. The method presented in this study can be used with any 2D image analysis software.

ACKNOWLEDGEMENTS

We thank A. Khelil and H. Bungartz for critically reading the manuscript. The research was supported by Grant Wu 268/1-2 from the German Research Foundation (DFG) to S.W. and by the Research Center for Fundamental Studies of Aerobic Biological Wastewater Treatement (SFB 411), Munich, Germany.

REFERENCES

Bloem, J., D.K. Van Mullem and P.R. Bolhuis. 1992. Microscopic counting and calculation of species abundance and statistics in real time with an MS-DOS personal computer, applied to bacteria in soil smears. *J. Microbiol. Methods* 16, 203–213.

Caldwell, D.E., D.R. Korber and J.R. Lawrence. 1992a. Confocal laser scanning microscopy and digital image analysis in microbial ecology. *Adv. Microbiol. Ecol.* 12, 1–67.

Caldwell, D.E., D.R. Korber and J.R. Lawrence. 1992b. Imaging of bacterial cells by fluorescence exclusion using scanning confocal laser microscopy. *J. Microbiol. Methods* 15, 249–261.

Kuehn, M., M. Hausner, H.-J. Bungartz, M. Wagner, P. A. Wilderer and S. Wuertz. 1998. Automated confocal laser scanning microscopy and semiautomated image processing for analysis of biofilms. *Appl. Environ. Microbiol.* 64, 4115–4127.

Lawrence, J.R., D.R. Korber and D.E. Caldwell. 1989. Computer-enhanced darkfield microscopy for the quantitative analysis of bacterial growth and behavior on surfaces. *J. Microbiol. Methods* 10, 123–138.

Lawrence, J.R., D.R. Korber, B.D. Hoyle, J.W. Costerton and D.E. Caldwell. 1991. Optical sectioning of biofilms. *J. Bacteriol.* 173, 6558–6567.

Moller, S., C.S. Kristensen, L.R. Poulsen, J.M. Carstensen and S. Molin. 1995. Bacterial growth on surfaces: automated image analysis for quantification of growth rate-related parameters. *Appl. Environ.Microbiol.* 61, 741–748.

CHAPTER 13

Confocal Laser Scanning Microscopy (CLSM) of Biofilms

THOMAS R. NEU

INTRODUCTION

B IOFILM systems, by definition, are complex microbial communities, and their extracellular products are associated with interfaces. Biofilms harbor a variety of organisms including bacteria, protozoa, algae, fungi, and viruses and metazoa such as nematodes or rotatoria. Furthermore, they accumulate non-living material, e.g., mineral particles, elements, detritus, humic substances, colloidal material, nutrients and contaminants (Arvin, 1995). This very diverse mix of compounds shows a large heterogeneity that resulted in an evolution of models trying to describe the complex architecture (Costerton et al., 1994; Keevil et al., 1995; Neu and Lawrence, 1997; Characklis and Wilderer, 1989; Wimpenny and Colasanti, 1997).

The biofilm structure is held together and attached to the interface by the extracellular polymeric substances (EPS) produced by the biofilm organisms (Christensen, 1989; Geesey, 1982; Neu and Marshall, 1990; Neu, 1992; Neu, 1994; Sutherland, 1983; Sutherland, 1995). EPS may be polysaccharides, proteins, nucleic acids or (other) amphiphilic polymers (Neu, 1996). All of these polymers have a high affinity for water. Therefore, the aqueous medium in which the biofilm structure is embedded represents a key factor for transport processes to, from and within the biofilm structure. Thus, biofilms are highly hydrated structures that may have a water content of up to 98% (Christensen and Characklis, 1990). The high water content eventually will determine the proper technique to microscopically examine the natural three-dimensional (3D) hydrated biofilm architecture.

The best approach by which biofilms may be probed should include two strategies: one technique able to give information about physiological processes, e.g., by using microsensors (Klimant et al., 1995; Revsbech and Jörgensen,

1986) or reporter genes (Prosser, 1994; Stewart and Williams, 1992) and a second that gives spatial information, e.g., microscopy. To visualize the biofilm structure, there are several techniques available (Marshall, 1986; Surman et al., 1996). Basically, the biofilm can be analyzed in three different ways: dehydrated, partly dehydrated and hydrated.

Dehydration is a prerequisite if electron microscopic techniques are used. For transmission electron microscopy (TEM), the sample has to be fixed, embedded, dehydrated, stained and sectioned. The information received will be cross-sectional including cell location within the biofilm and internal cell structures (Costerton, 1985). Scanning electron microscopy (SEM) also requires fixation and dehydration to visualize the sample. It will allow a 3D presentation of biofilm surface structures and the imaging of complex shapes (Richard and Turner, 1984; Sieburth et al., 1974). TEM and SEM have the problem of shrinking due to the necessity of dehydration and fixation of samples. For example, the EPS matrix in biofilms will be condensed to a fraction of its original volume. This fact means a loss of 3D information that eventually will make interpretation difficult. However, both TEM and SEM will allow high magnification not achieved by light microscopy. In addition, there is the option of elemental microanalysis with both electron microscopic techniques. Newer SEM systems are equipped with a cryo-transfer chamber and a cold stage for faster preparation and preservation of fine structures (Chenu and Jaunet, 1992; Neu and Marshall, 1991; Richards and Turner, 1984).

Partly dehydrated biofilms can be examined by environmental scanning electron microscopy (ESEM). With this technique moist samples may be used whereby shrinking artifacts can be minimized (Sutton et al., 1994). The result will give information closer to the real hydrated situation as has been shown for marine aggregates (Lavoie et al., 1995) and for streptococcal biofilms (Sutton et al., 1994). However, the electron beam will still damage the biological sample after a short time. Another technique for which partly dehydrated samples can be used is atomic force microscopy (AFM). The tip of the probe will give information on the surface contours of the sample. AFM will reveal a high-resolution image of the topography of single cells and biofilms (Bremer et al., 1992; Kasas et al., 1994). Nevertheless, AFM cannot elucidate internal biofilm structures.

Fully hydrated biofilms can be examined via light microscopic techniques. Normal light microscopy together with all its variations will be limited to samples being just a few micrometers thick. Because biofilms may be several hundred micrometers thick, new techniques are necessary to visualize the 3D structure of biofilms, e.g. confocal laser scanning microcopy (CLSM).

CLSM HISTORY

CLSM is associated with its inventor Marvin Minski who, at MIT in Cambridge, MA, in 1957, built the very first instrument (Minski, 1988). Despite a patent, nobody became interested in this technique. Ten years later Egger and Petran published an article on optical sectioning (Egger and Petran, 1967), but it took another 10 years until a group associated with Sheppard and Wilson in Oxford delivered the theoretical background for the understanding of confocal microscopy (summarized in Wilson and Sheppard, 1984). The development of CLSM was closely associated with inventions in several key technologies such as optics, lasers, fluorescence markers, detectors, computers and digital image analysis systems. The first instruments became commercially available in 1987 (Inoue, 1995). At present, more than 10 companies offer a range of different instruments (Pawley, 1995b). Initially, CLSM was used for various purposes in the medical field (Shotton, 1993). The first application of CLSM for the study of microbial biofilms was reported by Lawrence and coworkers in 1991 (Lawrence et al., 1991).

CONFOCAL IMAGING

What does confocal mean? It simply stands for the fact that visual information from structures that are not in focus will be eliminated and will not contribute to the final image. This is achieved in two ways: first, by illuminating only one point or one line and subsequent scanning of the defined focal planes, and second, by removing out-of-focus light with a detector pinhole. A complete CLSM system consists of several hardware and software components (Table 1). An overview of the various CLSM types currently on the market and different CLSM imaging techniques may be found in the "handbook" of confocal microscopy (Pawley, 1995a).

ADVANTAGES/DISADVANTAGES

The key advantage of CLSM is the optical sectioning of thick biological structures with no out-of-focus disturbance (Lichtman, 1994; Shotton and White, 1989). Since the first application of CLSM on biofilms, several groups

TABLE 1. Components of Complete CLSM System.

(1) A conventional normal or inverted epifluorescence microscope
(2) One or several lasers as a defined light source together with the necessary cooling system
(3) Scanning device for exact positioning of the sample
(4) Detector system, which is usually a photomultiplier
(5) A computer and/or workstation for controlling the system
(6) Two monitors one for the software and one with high resolution for the image
(7) Image analyzing system for quantification of 3D data sets
(8) Printing devices for direct video prints and/or for high-resolution digital printing
(9) A file system for archiving the huge 3D data sets

TABLE 2. Advantages of CLSM.

(1) Investigation of fully hydrated, living microbial communities up to several hundred micrometers thick
(2) Capability of noninvasive optical sectioning with virtually no out-of-focus blur
(3) Possibility of horizontal (*xy*), vertical (*xz*) and temporal (*xt*) sectioning
(4) Availability of numerous fluorescent (fluorescent mode) and non-fluorescent (reflection mode) probes
(5) Simultaneous application of multiple probes and simultaneous multichannel (4) recording of digitally enhanced signals
(6) Quantitative static and dynamic analyses of 3D organization of macromolecules, single cells and microcolonies
(7) 3D tomography, multicolor stereoscoptic imaging and computer animation

have shown the value of confocal microscopy as a routine tool in biofilm research. The advantages of CLSM are summarized in Table 2.

Nevertheless, CLSM is not without disadvantages. The first to be encountered is the cost of buying a system. The above-mentioned, complete CLSM system will comprise an investment of nearly half a million $ U.S. Another disadvantage is the considerable time required for scanning. This means during scanning the computer is busy and not able to carry out any other task. Another consequence of the prolonged scanning time is that only non-motile organisms can be scanned. This is no problem with most biofilms as the cells are immobilized. However, some bacteria and also protozoa do move within biofilms and may result in unwanted tracks in the image. A further problem can be caused by the fluorescent stains used, because they may have a negative effect on the living organisms. This may not be the case with most bacteria, but for example, protozoa are quite sensitive to some of the stains. An even greater problem may be bleaching of fluorochromes if quantification is desired. There are several solutions to this disadvantage, e.g., selecting the "right fluorochrome," application of antifading agents, correct adjustment of laser power and minimization of sample illumination. Bleaching has been partly solved because of the development of new fluorochromes that are more photostable such as BODIPY or the CY stains. Finally, the optical system of a CLSM causes problems because it is far from ideal: the objective lenses are designed for operation under very restricted conditions; also, the biological sample does not represent an ideal object for optical examination (Laurent et al., 1994).

TECHNIQUES

PROCEDURE

CLSM is performed in three steps, all of which are crucial for an optimal result. The first step is the preparation and staining of the sample in the laboratory. Already at this stage one has to consider the properties of the sample, possible autofluorescence, the excitation and emission maxima of the chosen stain, the type of microscope (normal/inverted), the way in which the sample will be mounted, etc. The second step is the actual scanning of the sample. Most crucial is the correct alignment of the laser. Again, it is important to select the right settings for recording the data set. Depending on what is to be demonstrated with these data, it is important to be aware of type of lens, filter settings, pinhole, resolution, number of scans, averaging procedure, etc. The third step is digital image analysis of the 3D data set. What can be analyzed will depend on the proper selection of the parameters during preparation and scanning. This may determine the final result for presentation and quantification of image data (Lawrence et al., 1998a).

REFLECTION

Without any fluorescent stain, the CLSM can be run in reflection mode. Either the sample itself shows enough reflective signal to record the image or the sample may be "stained" with reflective particles. For this purpose the whole range of colloidal gold labeled compounds may be applied.

LASER SYSTEMS

The most common laser supplied with CLSM systems is the argon-krypton laser. This laser has three exitation lines at 488, 568 and 647 nm. As a consequence, triple labeling, e.g., using FITC, TRITC and CY5, becomes possible.

The UV laser is an expensive option that allows the use of the complete spectrum of visible light for CLSM. Despite the high price tag of an UV laser, it is an important tool for the application of blue fluorophores. Examples of these probes are routine stains such as DAPI for bacterial cell number, Hoechst 2495 for proteins, Calcofluor White for certain types of polysaccharides or AMCA labeled conjugates.

FLUORESCENT STAINS

Fluorescent stains have been limited in numbers in the past. Nowadays, there seems to be a continuous development of new stains. The range of molecules and their applications may be found in the handbook of a company currently dominating the market of fluorescent stains (Haugland, 1997). Table 3 gives an overview of the types of probes available at present. All together, there are probably more than 2000 different fluorescent compounds available which may be used for CLSM of biological samples. The limitations are only set by the range of wavelengths and appropriate filter sets in the visible part

TABLE 3. Types of Fluorescent Probes Available for CLSM.

Probes/Target	Examples of Fluorescent Stains
Proteins	Hoechst 2495
Nucleic acids	DAPI
	Hoechst 33258
	Syto
Polysaccharides	Clacofluor White
	FITC/TRITC lectins
Cell vitality	Live/dead BacLight
	Live/dead yeast
Bacterial cell type	Baclight gram
Cell organelles	Paclitaxel (tubulin)
	Mitotracker (mitochondria)
	Lysosensor (lysosomes)
Endocytosis	Range of fluorescent particles
Membrance probes	Range fluorescent lipids, fatty acids and sterols
Enzyme substrates	Fluorescein digalactoside
	4-Methylumbelliferyl phosphate
Receptors	Fluorescent œ-bungarotoxin
Transport inhibitors	Probes for ion channels
pH indicators	SNARF
Ion indicators	BAPTA
	Mag-Fura
Dextranes	Range of conjugates with various molecular weights
Micro-particles	Modified microspheres
Caged probes	Photoactivation of fluorescence in living cells
Antibodies	Polyclonal or monoclonal against any structure of interest
Genes	Numerous types of gene probes for phylogenetic and structural genes

of the spectrum. Despite this limitation, multi-color labeling using up to 7 probes (Amann et al., 1996; Ried et al., 1992) and even 24 colors (Schröck et al., 1996) has been demonstrated.

MULTICHANNEL

The use of more than one fluorescent stain may require sequential scans of the various signals. New instruments are capable of simultaneously recording up to four channels, which means a sample may be stained with up to four different stains leading to an image with four times the information. Examples may be the use of several types of gene probes, use of nucleic acid probes in combination with the fluorescence of certain contaminants, simultaneous application of gene probes and antibodies and the combination of cell probes with probes for macromolecules (see the "Applications" section).

Further information may be obtained from other reviews discussing the various aspects of CLSM and digital image analysis applied in environmental microbiology and biofilm research (Caldwell et al., 1992; Lawrence and Hendry, 1996; Lawrence et al., 1998a).

APPLICATIONS

Most CLSM pioneering work in microbiology originated from Lawrence and coworkers. They demonstrated the applicability of optical sectioning of fully hydrated, living pure culture biofilms. The *xy* and *xz* sections showed the open biofilm structure that was composed of extracellular polymeric material and void spaces (Lawrence et al., 1991). The technique of negative staining of bacterial cells was later described in detail (Caldwell et al., 1992). They also showed the different behavior of motile and non-motile *Pseudomonas fluorescens* cells during adhesion and biofilm development (Korber et al., 1993; Korber et al., 1994). As a consequence of biofilm heterogeneity they measured diffusion coefficients in biofilms by using fluorescence recovery after photobleaching (FRAP) in conjunction with CLSM (Lawrence et al., 1994). In a later step they studied a six-member biofilm community able to degrade chlorinated organic compounds. They were able to show spatial relationships between the members of the biofilm community (Wolfaardt et al., 1994). In addition, the adsorption and accumulation of the organic contaminant in the EPS matrix was found (Wolfaardt et al., 1995). Finally, they investigated complex biofilm communities grown with river water. In this study new structural features of biofilms in natural habitats were described (Neu and Lawrence, 1997). A different biofilm research field is represented by bacteria found in geologic media (Lawrence and Hendry, 1996). For *Thiobacillus* spp., the selective colonization of sulfide minerals could be shown by using viable negative and positive staining in combination with CLSM. The cells preferentially adhered to the most electrochemically active mineral, e.g., pyrrhotite (Lawrence et al., 1997a, 1997b). Recently, the simultaneous or sequential collection of multiple data in three channels has been reported for microbial biofilms. The study demonstrated the possibility to collect the cellular signal (nucleic acid stain), the polysaccharide signal (lectin binding) and the autofluorescent signal of photosynthetic organisms (Lawrence et al., 1998b). A similar approach was used for in situ localization of exopolymers in a defined multispecies biofilm. In this study the presence of chlorinated organics could be correlated with the distribution of certain exopolymers (Wolfaardt et al., 1998).

In situ hybridization of bacteria with fluorescence-labeled rRNA-directed oligonucleotides (FISH) is another approach that takes advantage of CLSM. It is ideally suited to record the signal from individual cells within complex microbial communities (Amann et al., 1995). By using CLSM the spatial relationships of different bacterial groups within natural biofilm systems could be shown (Table 4). Very recently, microscopic localization of bacterial genetic material by bacterial chromosomal painting was reported. With this technique not only the phylogenetic relationship but also the metabolic and genetic potential of bacteria can be determined (Lanoil and Giovannoni, 1997).

A persistent disadvantage of in situ hybridization is the necessity for fixation with ethanol or paraformaldehyde. The fixation step is needed to enhance the permeability of the cell for the gene probes. Thereby the 3D information is partly lost due to shrinking of the biofilm. In natural complex systems containing humic and detrital compounds this may not be as dramatic as in pure culture systems comprising only bacterial cells and EPS. Nevertheless, there is an urgent need to make cells permeable for gene probes without changing the biofilm architecture. Some groups used embedding in agarose to preserve the biofilm structure prior to staining with fluorophores (Massol-Deya et al., 1995; Möller et al., 1996). However, to the author's knowledge only one report on embedding and subsequent in situ hybridization has been published (Macnaughton et al., 1996).

Other research groups have also succesfully used CLSM to study a variety of aspects in environmental and interfacial microbiology (Ghiorse et al., 1996). The now generally accepted structural feature of channels within biofilms has also been described in fixed film reactors treating petroleum-contaminated groundwater (Massol-Deya et al., 1995). CLSM was used to quantitatively study the germination of individual *Bacillus cereus* spores by following the change in refractility (Coote et al., 1995). For routine determinations, a fully automatic system to measure bacterial numbers, cell volumes and cell division was suggested. The image analyses of CLSM results was tested in the difficult matrix of soil smears with reasonable agreement compared with other methods (Bloem et al., 1995). The distribution and activity of *E. coli* immobilized in a thin synthetic matrix have been reported. The method was tested by using CLSM to assess the value of longterm biocatalysis (Swope and Flickinger, 1996). Quantification of plankton and detritus is a central issue in aquatic ecology. By using a set of fluorophores in combination with CLSM it was possible to separately calculate the pool of plankton and detritus in aquatic ecosystems (Verity et al., 1996). Diffusion coefficients are an important measure for studying mass transport within biofilms. Microinjection of fluorescent dyes with subsequent quantification using CLSM has been described. With this technique the cell cluster matrix was calculated as a gel with 0.6-nm fibers and 80-nm pores (Beer et al., 1997). The coexistence of two species in a laboratory-grown biofilm has been investigated and examined by CLSM. It was hypothesized that microscale heterogeneity and differential adhesion and de-adhesion were responsible for the coexistence (Stewart et al., 1997). Corrosion inhibition of aerobic biofilms was studied by using CLSM. It has been suggested that the protective effect of *Pseudomonas fragi* or *E. coli* is caused by a live biofilm and its metabolism (Jayaraman et al., 1997). Detection, identification and enumeration of pathogenic agents, such as *Cryptosporidium parvum* oocysts, is another topic in environmental microbiology. For this purpose a fluorogenic stain was combined with an immunostaining procedure to confidently identify the oocysts in soil, sediment and feces. Comparison of color epifluorescence, differential interference contrast and CLSM greatly facilitated the interpretation of results (Anguish and Ghiorse, 1997). CLSM can be also used to study plant pathogens on leaf surfaces. So far, only SEM has been used to investigate the phyllosphere. CLSM may reveal further details of the interfacial community on leaves (Morris et al., 1997).

CONCLUSIONS

CLSM allows the recording of unmatched multidimensional data sets of fully hydrated living biofilm systems not possible with any other system. However, there are still several future needs that can be determined from the disadvantages of current CLSM systems. First, non-toxic fluorescent stains should be developed. In addition, new environmentally sensitive stains would be of interest to probe for the microhabitats within biofilms. Second, new objective lenses that are corrected for confocal laser scanning microscopy are needed. Scanning time is one area that needs improved. Ideally, it would be an desirable to acquire an immediate scan in real time to be useful if motility is of interest. The costs of the CLSM systems will decrease if computer hardware becomes cheaper. Nevertheless, there is a need for software that is able to quantitate the 3D data sets. With new inventions CLSM and its variations, e.g., *xy/xz* and *xt*-dimensions, reflection mode, simultaneous multichannel scanning, fluorescence recovery after photobleaching (FRAP), photoactivation of fluorescence (PAF), color spectroscopy and Raman spectroscopy, have a large potential in quantitative microscopy. These future potentials will finally lead to a significant increase in our knowledge on the structure and function of biofilm systems.

ACKNOWLEDGEMENT

The author is indebted to Ute Kuhlicke for retyping the manuscript.

REFERENCES

Anguish, L.J. and W.C. Ghiorse. 1997. Computer assisted laser scanning and video microscopy for analysis of *Cryptosporidium parvum* oocysts in soil, sediment, and feces. *Appl. Environ. Microbiol.* 63, 724–733.

Amann, R.I., W. Ludwig and K.-H. Schleifer. 1995. Phylogenetic identification and in situ detection of individual microbial cells without cultivation. *Microbiol. Rev.* 59, 143–169.

Amann, R., J. Snaidr, M. Wagner, W. Ludwig and K.-H. Schleifer. 1996. In situ visualization of high genetic diversity in a natural microbial community. *J. Bacteriol.* 178, 3496–3500.

Arvin, E. 1995. Biological degradation of organic chemical pollutants in biofilm systems. *Water Sci. Technol.* 31, 1–280.

Assmuss, B., M. Schloter, G. Kirchhof, P. Hutzler and A. Hartmann. 1997. Improved in situ tracking of rhizosphere bacteria using dual staining with fluorescence-labeled antibodies and rRNA-targeted oligonucleotide probes. *Microbiol. Ecol.* 33, 32–40.

Assmuss, B., P. Hutzler, G. Kirchhof, R. Amann, J.R. Lawrence and A. Hartmann. 1995. In situ localization of *Azospirillum brasilense* in the rhizosphere of wheat with fluorescently labeled, rRNA-targeted oligonucleotide probes and scanning confocal laser microscopy. *Appl. Environ. Microbiol.* 61, 1013–1019.

Beer, D.de, P. Stoodley and Z. Lewandowski. 1997. Measurement of local diffusion coefficients in biofilms by microinjection and confocal microscopy. *Biotechnol. Bioeng.* 53, 151–158.

Bloem, J., M. Veninga and J. Shepard. 1995. Fully automatic determination of soil bacterium numbers, cell volumes, and frequencies of dividing cells by confocal laser scanning microscopy and image analysis. *Appl. Environ. Microbiol.* 61, 926–936.

Bremer, P.J., G.G. Geesey and B. Drake. 1992. Atomic force microscopy examination of the topography of a hydrated bacterial biofilm on a copper surface. *Curr. Microbiol.* 24, 223–230.

Caldwell, D.E., D.R. Korber and J.R. Lawrence. 1992. Imaging of bacterial cells by fluorescence exclusion using scanning confocal laser microscopy. *J. Microbiol. Methods* 15, 249–261.

Characklis, W.G. and P.A. Wilderer. 1989. *Structure and function of biofilms.* Dahlem workshop report 46. John Wiley & Sons, Chichester.

Chenu, C. and A.M. Jaunet. 1992. Cryoscanning electron microscopy of microbial extracellular polysaccharides and their association with minerals. *Scanning* 14, 360–364.

Christensen, B.E. 1989. The role of extracellular polysaccharides in biofilms. *J. Biotechnol.* 10, 181–202.

Christensen, B.E. and W.G. Characklis. 1990. Physical and chemical properties of biofilms. In: W.G. Characklis and K.C. Marshall (eds.): Biofilms. John Wiley & Sons, New York, 93–130.

Coote, P.J., C.M.-P. Billon, S. Pennell, P.J. McClure, D.P. Ferdinando and M.B. Cole. 1995. The use of confocal scanning laser microscopy (CLSM) to study the germination of individual spores of Bacillus cereus. *J. Microbiol. Methods* 21, 193–208.

Costerton, J.W. 1985. The role of bacterial exopolysaccharides in nature and disease. *Dev. Ind. Microbiol.* 26, 249–261.

Costerton, J.W., Z. Lewandowski, D. DeBeer, D. Caldwell, D. Korber and G. James. 1994. Biofilms, the customized microniche. *J. Bacteriol.* 176, 2137–2142.

Egger, M.D. and M. Petran. 1967. New reflected-light microscope for viewing unstained brain and ganglion cells. *Science* 157, 305–307.

Geesey, G.G. 1982. Microbial exopolymers: ecological and economic considerations. *ASN News* 48, 9–14.

Ghiorse, W.C., D.N. Miller, R.L. Sandoli and P.L. Siering. 1996. Applications of laser scanning microscopy for analysis of aquatic microhabitats. *Microsc. Tes. Tech.* 33, 73–86.

Haugland, R.P. 1997. *Handbook of fluorescent probes and research chemicals.* Molecular Probes, Eugene (OR).

Inoue, S. 1995. Foundations of confocal scanned imaging in light microscopy. In: J.B. Pawley (ed.): *Handbook of biological confocal microscopy.* Plenum Press, New York.

Jayaraman, A., J.C. Earthman and T.K. Wood. 1997. Corrosion inhibition by aerobic biofilms on SAE 1018 steel. *Appl. Microbiol. Biotechnol.* 47, 62–68.

Kasas, S., B. Fellay and R. Cargnello. 1994. Observation of the action of penicillin on *Bacillus subtilis* using atomic force microscopy: technique for the preparation of bacteria. *Surface Interface Anal.* 21, 400–401.

Keevil, C.W., J. Rogers and J.T. Walker. 1995. Potable water biofilms. *Microbiol. Eur.* 3, 10–14.

Klimant, I., V. Meyer and M. Kühl. 1995. Fiber-optic oxygen microsensors, a new tool in aquatic biology. *Limnol. Oceanogr.* 40, 1159–1165

Korber, D.R., J.R. Lawrence and D.E. Caldwell. 1994. Effect of motility on surface colonization and reproductive success of *Pseudomonas fluorescens* in dual-dilution continuous culture and batch culture systems. *Appl. Environ. Microbiol.* 60, 1421–1429.

Korber, D.R., J.R. Lawrence, M.J. Hendry and D.E. Caldwell. 1993. Analysis of spatial variability within mot$^+$ and mot$^-$ *Pseudomonas fluorescens* biofilms using representative elements. *Biofouling* 7, 339–358.

Lanoil, B.D. and S.J. Giovannoni. 1997. Identification of bacterial cells by chromosomal painting. *Appl. Environ. Microbiol.* 63, 1118–1123.

Laurent, M., G. Johannin, N. Gilbert, L. Lucas, D. Cassio, P.X. Petit and A. Fleury. 1994. Power and limits of laser scanning confocal microscopy. *Biol. Cell.* 80, 229–240.

Lavoie, D.M., B.J. Little, R.I. Ray, R.H. Bennett, M.W. Lambert, V. Asper and R.J. Baerwalds. 1995. Environmental scanning electron microscopy of marine aggregates. *J. Microscopy* 178, 101–106.

Lawrence, J.R. and M.J. Hendry. 1996. Transport of bacteria through geologic media. *Can. J. Microbiol.* 42, 410–422.

Lawrence, J.R., G.M. Wolfaardt and D.R. Korber. 1994. Determination of diffusion coefficients in biofilms by confocal laser microscopy. *Appl. Environ. Microbiol.* 60, 1166–1173.

Lawrence, J.R., D.R. Korber, G.M. Wolfaardt and D.E. Caldwell. 1997a. Analytical imaging and microscopy techniques. In: C.J. Hurst, G.R. Knudson, M.J. McInerney, L.D. Stetzenbach and M.V. Walker (eds.): *Manual for environmental microbiology.* American Society for Microbiology, NY, 29–51.

Lawrence, J.R., Y.T.J. Kwong and G.D.W. Swerhone. 1997b. Colonization and weathering of natural sulfide mineral assemblages by *Thiobacillus ferrooxidans. Can. J. Microbiol.* 43, 178–188.

Lawrence, J.R., G.M. Wolfaardt and T.R. Neu. 1998a. The study of biofilms using confocal laser scanning microscopy. In: M.H.F. Wilkinson and F. Shut (eds.): *Digital image analysis of microbes.* John Wiley & Sons, New York, 431–465.

Lawrence, J.R., T.R. Neu and G.D.W. Swerhone. 1998b. Application of multiple parameter imaging for the quantification of algal, bacterial and exopolymer components of microbial biofilms. *J. Microbiol. Methods* 32, 253–261.

Lawrence, J.R., D.R. Korber, B.D. Hoyle, J.W. Costerton and D.E. Caldwell. 1991. Optical sectioning of microbial biofilms. *J. Bacteriol.* 173, 6558–6567.

Lichtman, J.W. 1994. Confocal microscopy. *Scientific American* October, 78–84.

Macnaughton, S.J., T. Booth, T.M. Embley and A.G. O'Donnell. 1996. Physical stabilization and confocal microscopy of bacteria on roots using 16s rRNA targeted, fluorescent-labeled oligonucleotide probes. *J. Microbiol. Methods* 26, 279–285.

Manz, W., M. Eisenbrecher, T. R. Neu and U. Szewzyk. 1998. Abundance and spatial organization of gram-negative sulfate-reducing bacteria in activated sludge investigated by in situ probing with specific 16S rRNA targeted oligonucleotides. *FEMS Microbiol. Ecol.* 25, 43–61.

Marshall, K.C. 1986. Microscopic methods for the study of bacterial behaviour at inert surfaces. *J. Microbiol. Methods* 4, 217–227.

Massol-Deya, A.A., J. Whallon, R.F. Hickey and J.M. Tiedje. 1995. Channel structures in aerobic biofilms of fixed-film reactors treating contaminated groundwater. *Appl. Environ Microbiol.* 61, 769–777.

Minski, M. 1988. Memoir on inventing the confocal scanning microscope. *Scanning* 10, 128–138.

Möller, S., A.R. Pedersen, L.K. Poulsen, E. Arvin and S. Molin. 1996. Activity and three-dimensional distribution of toluene-degrading *Pseudomonas putida* in a multispecies biofilm assessed by quantitative in situ hybridization and scanning confocal laser microscopy. *Appl. Environ. Microbiol.* 62, 4632–4640.

Morris, C.E., J.-M. Monier and M.-A. Jaques. 1997. Methods for observing microbial biofilms directly on leaf surfaces and recovering them for isolation of culturable microorganisms. *Appl. Environ. Microbiol.* 63, 1570–1576.

Neu, T.R. 1992. Polysaccharide in biofilmen. In: P. Präve, M. Schlingmann, K. Esser, R. Thauer and F. Wagner (eds.): Jahrbuch Biotechnologie Band 4. Carl Hanser, München, 73–101.

Neu T.R. 1994. The challenge to analyse extracellular polymers in biofilms. In: L. Stal and P. Caumette (eds.): *Microbial mats*. Springer Verlag, Berlin, 221–227.

Neu T.R. 1996. Significance of bacterial surface-active compounds in interaction of bacteria with interfaces. *Microbiol. Rev.* 60, 151–166.

Neu, T.R. and K.C. Marshall. 1990. Bacterial polymers: physicochemical aspects of their interactions at interfaces. *J. Biomat. Appl.* 5, 107–133.

Neu, T.R. and K.C. Marshall. 1991. Microbial "footprints"— a new approach to adhesive polymers. *Biofouling* 3, 101–112.

Neu, T.R. and J.R. Lawrence. 1997. Development and structure of microbial biofilms in river water studied by confocal laser scanning microcopy. *FEMS Microbiol. Ecol.* 24, 11–25.

Pawley, J.B. (ed.). 1995a. *Handbook of biological confocal microscopy*. Plenum Press, New York.

Pawley, J.B. 1995b. Appendix 2: Light paths of current commericial confocal microscopes for biology. In: J.B. Pawley (ed.): *Handbook of biological confocal microscopy*. Plenum Press, New York, 581–598.

Prosser, J.I. 1994. Molecular marker systems for detection of genetically engineered micro-organisms in the environment. *Microbiology* 140, 5–17.

Revsbech, N.P. and B.B. Jörgensen. 1986. Microelectrodes: their use in microbial ecology. *Adv. Microbiol. Ecol.* 9, 293–352.

Richard, S.R. and R.J. Turner. 1984. A comparative study of techniques for the examination of biofilms by scanning electron microscopy. *Water Res.* 18, 767–773.

Ried, T., A. Baldini, T.C. Rand and D.C. Ward. 1992. Simultaneous visualization of seven different DNA probes by in situ hybridization using combinatorial fluorescence and digital imaging microscopy. *Proc. Natl. Acad. Sci. USA* 89, 1388–1392.

Rothemund, C., R. Amann, S. Klugbauer, W. Manz, C. Bieber, K.-H. Schleifer and P. Wilderer. 1996. Microflora of 2,4-dichlorophenoxyacetic acid degrading biofilms on gas permeable membranes. *Syt. Appl. Microbiol.* 19, 608–615.

Schloter, M., R. Borlinghaus, W. Bode and A. Hartmann. 1993. Direct identification, and localization of *Azospirillum* in the rhizosphere of wheat using fluorescence-labelled monoclonal antibodies and confocal scanning laser microscopy. *J. Microsc.* 171, 173–177.

Schramm, A., L.H. Larsen, N.P. Revsbech, N.B. Ramsing, R. Amann and K.-H. Schleifer. 1996. Structure and function of a nitrifying biofilm as determined by in situ hybridization and the use of microelectrodes. *Appl. Environ. Microbiol.* 62, 4641–4647.

Schröck, E., S. du Manoir, T. Veldman, B. Schoell, J. Wienberg, M.A. Ferguson-Smith, Y. Ning, D.H. Ledbetter, D. Bar-Am, Y. Garinin and T. Ried. 1996. Multicolour spectral karyotyping of human chromosomes. *Science* 273, 494–497.

Shotton, D. (ed.). 1993. *Electronic light microscopy—techniques in modern biomedical microscopy.* Wiley, New York.

Shotton, D. and White, N. 1989. Confocal scanning microscopy: three-dimensional biological imaging. *Trends Biochem. Sci.* 14, 435–439.

Sieburth, J.McN., R.D. Brooks, R.V. Gessner, C.D. Thomas and J.L. Tootle. 1974. Microbial colonization of marine plant surfaces as observed by scanning electron microscopy. In: R.R. Collwell and R.Y. Morita (eds): *Effect of the ocean environment on microbial activities.* University Park Press, Baltimore, 418–432.

Siering, P.L. and W.C. Ghiorse. 1997. Development and application of 16S rRNA-targeted probes for detection of iron- and manganese-oxidizing sheathed bacteria in environmental samples. *Appl. Environ. Microbiol.* 63, 644–651.

Stewart, G.S.A.B. and P. Williams. 1992. Lux genes and the applications of bacterial bioluminescence. *J. Gen. Microbiol.* 138, 1289–1300.

Stewart, P.S., A.K. Camper, S.D. Handran, C.-T. Huang and M. Warnecke. 1997. Spatial distribution and coexistence of *Klebsiella pneumoniae* and *Pseudomonas aeruginosa* in biofilms. *Microbiol. Ecol.* 33, 2–10.

Surman, S.B., J.T. Walker, D.T. Goddard, L.H.G. Morton, C.W. Keevil, W. Weaver, A. Skinner, K. Hanson, D. Caldwell and J. Kurtz. 1996. Comparison of microscope techniques for the examination of biofilms. *J. Microbiol. Methods* 25, 57–70.

Sutherland, I.W. 1983. Microbial polysaccharides—their role in microbial adhesion in aqueous systems. *CRC Crit. Rev. Microbiol.* 10, 173–201.

Sutherland, I.W. 1995. Biofilm-specific polysaccharides—do they exist? In: J. Wimpenny, P. Handley, P. Gilbert, H. Lappin-Scott (eds.): BioLine, Cardiff, 103–106.

Sutton, N.A., N. Hughes and P. Handley. 1994. A comparison of conventional SEM techniques, low temperature SEM and the electroscan wet scanning electron microscope to study the structure of a biofilm of *Streptococcus crista* CR3. *J. Appl. Bacteriol.* 76, 448–454.

Swope, K.L. and M.C. Flickinger. 1996. The use of confocal scanning laser microscopy and other tools to characterize *Escherichia coli* in a high-cell-density synthetic biofilm. *Biotechnol. Bioeng.* 52, 340–356.

Verity, P.G., T.M. Beatty and S.C. Williams. 1996. Visualization and quantification of plankton and detritus using digital confocal microscopy. *Aquatic Microbiol. Ecol.* 10, 55–67.

Wagner M., R. Amann, P. Kämpfer, B. Assmuss, A. Hartmann, P. Hutzler, N. Springer and K.-H. Schleifer. 1994a. Identification and in situ detection of Gram-negative filamentous bacteria in activated sludge. *Sys. Appl. Microbiol.* 17, 405–417.

Wagner M., B. Assmuss, A. Hartmann, P. Hutzler and R. Amann. 1994b. In situ analysis of microbial consortia in activated sludge using fluorescently labelled, rRNA-targeted oligonucleotide probes and confocal scanning laser microscopy. *J. Microsc.* 176, 181–187.

Wilson, T. and C.J.R. Sheppard. 1984. *Theory and practice of scanning optical microscopy.* Academic Press, London.

Wimpenny, J.W.T. and R. Colasanti. 1997. A unifying hypothesis for the structure of microbial biofilms based on cellular automaton models. *FEMS Microbiol. Lett.* 22, 1–16.

Wolfaardt, G.M., J.R. Lawrence, R.D. Robarts and D.E. Caldwell. 1995. Bioaccumulation of the herbicide diclofop in extracellular polymers and its utilization by a biofilm community during starvation. *Appl. Environ. Microbiol.* 61, 152–158.

Wolfaardt, G.M., J.R. Lawrence, R.D. Robarts and D.E. Caldwell. 1998. In situ characterization of biofilm exopolymers involved in the accumulation of chlorinated organics. *Microbiol. Ecol.* 35, 213–223.

Wolfaardt, G.M., J.R. Lawrence, R.D. Robarts, S.J. Caldwell and D.E. Caldwell. 1994. Multicellular organization in a degradative biofilm community. *Appl. Environ. Microbiol.* 60, 434–446.

CHAPTER 14

Microelectrodes and Tube Reactors in Biofilm Research

HARALD HORN

INTRODUCTION

MICROELECTRODE studies in biofilms were published as early as 1969 by Bungay et al. (1969). Other researchers followed with detailed microelectrode studies in several types of biofilms. Kuenen et al. (1986) published studies of oxygen profiles in trickling filter biofilms from wastewater treatment. Their research focused on light-dark effects on oxygen concentration in the biofilm. In addition, they investigated the effect of flow velocity on mass transfer in the boundary layer at the bulk/biofilm interface (Kuenen et al., 1986). As the Clark-type oxygen microelectrode was improved, more results were published (Debus, 1993; Hooijmans, 1990; Horn, 1992; Lewandowski et al., 1993).

Furthermore, microelectrodes for new measurable solutes were presented (Centonze et al., 1992; Cronenberg and van den Heuvel, 1991; Revsbech et al., 1988; Sweerts and de Beer, 1989).

Nitrate gradients could be measured in the littoral and profundal sediments of a mesoeutrophic lake with an ion-selective nitrate microelectrode (Sweerts and de Beer, 1989). The diffusion coefficient for glucose could be determined in a biocatalyst gel bead by measuring glucose profiles with a glucose enzyme microelectrode (Cronenberg and van den Heuvel, 1991). With a combined N_2O/oxygen-microelectrode, the conditions for denitrification in trickling filter biofilms could be formulated in detail (Dalsgaard and Revsbech, 1992). The microelectrode technique has become an important tool in biofilm research. Because of its high sensitivity, the oxygen microelectrode is still the most used microelectrode in this research area. The purpose of this chapter is to show the fundamental properties of oxygen microelectrodes. Furthermore, the biofilm tube reactor that was constructed especially for microelectrode measurements under defined flow conditions is presented. Some results of recent studies are also presented to show the advantages of the microelectrode.

MATERIALS AND METHODS

THE CLARK-TYPE OXYGEN MICROELECTRODE

The principle of the Clark-type oxygen electrode is based on the reduction of oxygen at the cathode:

$$O_2 + 4e^- + 2H_2O \rightarrow 4OH^- \tag{1}$$

The necessary potential between the cathode and the Ag/AgCl anode is about 600–700 mV. Originally, oxygen needle microelectrodes combined with external reference electrodes were used when measuring oxygen profiles in sediments or biofilm systems (Horn, 1992; Lorenzen et al., 1995). Studies conducted with Clark-type oxygen microelectrodes have been presented mainly in the last 6 years (Fu et al., 1993; Glud et al., 1992; Horn and Hempel, 1995; Lens et al., 1995). The advantage of this electrode is its insensitivity to electric fields. Figure 1 shows a sketch of the Clark-type oxygen microelectrode as it was used for the results being presented.

The current was measured with a pico ammeter (Keithley 617). This ammeter has an integrated potential source. The signal could be analyzed via an IEEE 488 interface on a personal computer.

Two main aspects have to be taken into account for the construction of the oxygen microelectrode:

(1) The response time (t_{90}) and
(2) The so-called "stirring effect," which means the sensitivity to flow velocity in the bulk phase (Gust et al., 1987).

The two factors are interdependent. On the one hand, a thick membrane at the tip of the microelectrode reduces the effect caused by changing flow velocity in the bulk phase to a minimum, but, on the other hand, increases the response time to unacceptable values.

Usually, a membrane thickness of about 10–20 μm leads to acceptable results with respect to the response time and the stirring effect. Figure 2 shows the results of two oxygen microelectrodes with different relative membrane thicknesses. Electrode 1 has a membrane thickness of <10 μm, and electrode 2 has a membrane thickness of 20 μm. The electrodes were exposed to air and then dipped in nitrogen-saturated water and vice versa. The signal in air was set at 100%; the signal in the nitrogen-saturated water was set at zero. It could be seen that the response time increased from ~1 s to 3 s with increas-

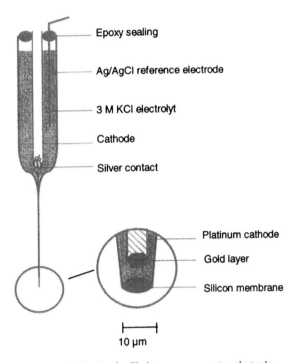

Epoxy sealing

Ag/AgCl reference electrode

3 M KCl electrolyt

Cathode

Silver contact

Platinum cathode

Gold layer

Silicon membrane

10 µm

Figure 1 Sketch of a Clark-type oxygen microelectrode.

ing membrane thickness. On the other hand, electrode 2 showed no "stirring effect." Figure 3 shows the current of electrode 2 in air-saturated water during the change of the flow velocity in a flow chamber. There was no appreciable effect on the microelectrode signal because of the change of the flow velocity, when the flow was set to zero.

This microelectrode could be used for the measurement of substrate conversion in the biofilm and mass transport in the bulk/biofilm interface. The experimental set up for such measurements is shown in Figure 4. The microelectrode was positioned above and in the biofilm by means of a micromanipulator. The micromanipulator enabled the positioning of the microelectrode with a precision of 10 µm in z-direction. Many oxygen profiles could be measured in this manner in a very short time. The shape of the oxygen profiles was dependent on the hydrodynamic conditions and the substrate loading during the growth phase of the biofilm. The two profiles indicated in Figure 4 give an idea of the possible oxygen profiles that can be measured in biofilms.

To gather more information for the complexity of growth and decay of microorganisms in biofilm systems, it was necessary to have defined conditions for flow and substrate loading during the whole development of the investigated biofilm. For this purpose a biofilm tube reactor was developed.

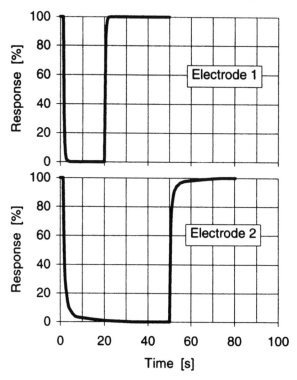

Figure 2 Response of two oxygen microelectrodes with different membrane thickness, electrode 1 (<10 μm) and electrode 2 (20 μm)

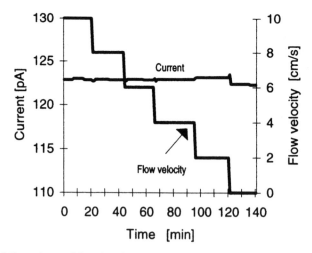

Figure 3 Dependence of the microelectrode current on the flow velocity in a flow chamber.

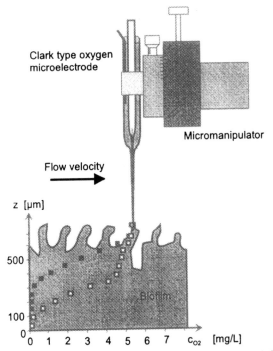

Figure 4 Experimental setup for microelectrode measurement in biofilms.

THE BIOFILM TUBE REACTOR

The biofilm tube reactor is shown in Figure 5. Flow velocity, substrate and oxygen conditions could be controlled independently. The ports for micro-electrodes allowed in situ measurement of oxygen profiles in the biofilm. To make the inner surface of the glass tube rougher, it was etched with hydrogen fluoride. This was performed to accelerate the growth of biofilms in the re-actor. The tube reactor could be darkened to avoid the growth of phototrophic microorganisms.

The biofilm thickness was measured by the method described by Charack-lis after draining the whole reactor for 30 min (Characklis, 1989). In addition, the mean biofilm thickness could be calculated from the mass of the wet biofilm. It was assumed that the density of the wet biofilm was 1 g cm^{-3}. The wet biofilm mass was determined by weighing the whole drained reactor. Both results corresponded very well (Horn and Hempel, 1997a). Small segments of 4–6 cm in length (Figure 5) allowed the determination of dry biomass (DM) per area substratum. After removing these segments from the reactor, drying them at 110°C and weighing them, the mean density (gm^{-3}) of the biofilm could be calculated for all phases of the investigation. The mean density of

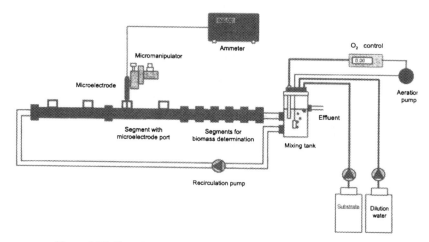

Figure 5 Biofilm tube reactor with setup for microelectrode measurement.

the biofilm relates to the entire biofilm volume (Horn and Hempel, 1997b). The segments of the tube reactor were connected by flanges and clamps. The specification of the biofilm tube reactor is shown in Table 1. Up to now it has been possible to investigate biofilms in the laminar region with a Reynolds number of up to 2000 and in regions of turbulent flow up to a Reynolds number of 5000. This allows the possibility to investigate the mass transfer, the detachment rate and the structure of the biofilm. The reactor could be modeled as a quasi-completely mixed reactor when the recirculation flow rate was set high enough. At recirculation flow rates above 0.04 L s^{-1}, there was no appreciable difference in the oxygen concentration under maximum substrate loading between the influent and effluent of the reactor. On the other hand, significant gradients of the substrate and the oxygen concentration could only be measured and obtained in laminar regions with a Reynolds number <1000.

The sampling of the grown biofilm for off-line analysis can be managed over a long investigation phase without problems. The short segments allow representative and convenient sampling of the biofilm. The reactor becomes shorter after sampling, but this does not present a problem, because the substrate loading can be adapted.

TABLE 1. Specification of the Biofilm Tube Reactor.

Parameter	Reactor without Biofilm
Reactor length	1.42 m
Tube diameter	0.026 m
Surface area	0.116 m^2
Reynolds number	100–5000

Apart from the above-described method for the determination of biofilm thickness, the possibility of applying optical methods for the monitoring of biofilm thickness is available. The operation under septic conditions is possible but has not been performed yet. The material of the reactor is not limited to glass. It was also possible to build the reactor out of metallic and plastic material.

The most important characteristic of the biofilm tube reactor is the possibility of measuring oxygen profiles in situ at different points in the tube An aperture of 4 mm diameter allowed the measurement of oxygen profiles in a small area opposite the electrode port. The hole was purposely made small as a concession to the hydrodynamic conditions, which would be influenced by bigger holes.

Figure 5 shows the setup for the measurement of oxygen profiles in the biofilm tube reactor.

RESULTS

MASS TRANSFER AT THE INTERFACE BULK/BIOFILM

The mass transfer at the interface bulk/biofilm is of great interest when modeling substrate conversion and mass transport in biofilm reactors and in environmental biofilms, for example, benthic biofilms in small running waters (Cazelles et al., 1991; Hickey, 1988; Leu et al., 1996). The substrate flux $j_{C,i}$ at the interface bulk/biofilm can be described in mathematical terms:

$$j_{C,i} = \beta(C_{Bi} - C_{Li}) \qquad (2)$$

C_{Bi} is the concentration of the substrate i in the bulk phase, and C_{Li} is the concentration at the biofilm surface. Because of the lack of mass transfer coefficients for biofilm systems in the past, the coefficient β was calculated by the equation:

$$\beta = D/L_L \qquad (3)$$

D is the diffusion coefficient and L_L is the thickness of the laminar sublayer, which can be calculated from hydrodynamic laws. Measurements with the oxygen microelectrode showed that the thickness of the concentration layer above the biofilm surface is smaller than the calculated laminar sublayer (Horn, 1992). Similar results have been achieved for biocatalyst beads (Hooijmans, 1990). Furthermore, the measured oxygen profiles show no linear behavior, as was assumed in equation (3). Nevertheless, the available data from microelectrode studies correspond well (Debus, 1993; Horn, 1992; Horn and Hempel, 1995;

Kuenen et al., 1986; Kühl et al., 1995; Lens et al., 1995; Lewandowski et al., 1993). Figure 6 shows the estimated thickness of the concentration layers based on the above-mentioned studies that have been conducted in the last 12 years. The estimates were made impressionistically. Despite the fact that most of the measurements were made in different reactors under various substrate conditions, the data correspond well. Especially in the region with flow velocities above 5 cm s^{-1}, the deviation is very small. This leads to the assumption that mass transfer in biofilm reactors with defined hydrodynamic conditions could be modeled on the basis of mass transfer coefficients calculated from microelectrode studies. Nevertheless, it has to be taken into account that the microelectrode may cause a change in the local flow pattern in the overlying water during the measurement. Lorenzen et al. stated that the concentration layer was suppressed by the tip of the microelectrode (Lorenzen et al., 1995). These measurements were made both from above and from under the biofilm. The concentration layer was found to be one-third smaller when measuring from under the biofilm than when applying the traditional measuring procedure from above. These results have to be verified by more measurements.

The results from the biofilm tube reactor (Figure 5) were used to formulate a relation for the Sherwood number Sh in tubes with biofilms:

$$Sh = 2Re^{1/2}Sc^{1/2}\ (d/L)^{1/2} * (1 + 0.0021\text{Re}) = \beta d/D \qquad (4)$$

Here, Sc is the Schmidt number, d is the tube diameter and L the reactor length. This equation can be used to determine the mass transfer coefficient

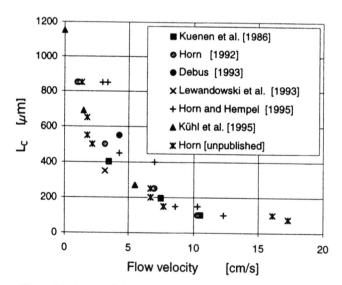

Figure 6 Thickness of the concentration layer above the biofilm surface.

for biofilms in reactors for wastewater treatment. The investigated biofilms were grown under substrate conditions similar to those in the effluent of primary wastewater treatment.

GROWTH AND DECAY IN AN AUTO-/HETEROTROPHIC BIOFILM

The main processes that dominate the performance of a biofilm are

- the transport of soluble and particulate materials in the biofilm
- the utilization of substrate for the process of growth of the microorganisms
- the decay of the microorganisms in the biofilm matrix

The microelectrode could be used to investigate both the substrate transport and the utilization of substrate by the microorganisms in the biofilm. When the substrate conditions were manipulated in a certain manner, the oxygen microelectrode was a very sensitive tool for the investigation of the growth and decay of autotrophic and heterotrophic microorganisms.

The following results show a biofilm that was grown for 140 days on an organic substrate (glucose) and then additionally loaded with ammonia for 64 days. The flow velocity in the tube reactor was 8 cm s^{-1} during the entire investigation. Table 2 shows the substrate loading and the oxygen concentration for all phases of the investigation. In phases I and II, only the heterotrophic microorganisms were supplied with their special substrate. Not until phase III did the autotrophic microorganisms get the chance to proliferate in the biofilm.

Attention was focused on the question whether the microelectrode measurement could be used to work out the two processes of growth and decay of the microorganisms. As can be seen in Table 2, between day 109 and day 140, there was no substrate supply. This phase was used to investigate the respiration of the biofilm and thereby the behavior of the microorganisms under non-substrate conditions, i.e., starvation.

Figure 7 shows two oxygen profiles in the starving biofilm on day 126. One oxygen profile was measured under normal flow conditions ($v = 8$ cm s^{-1}). This profile yielded no information due to the minimal oxygen consumption of the biofilm. On the other hand, the oxygen profile in a stagnant bulk fluid

TABLE 2. Substrate Loading and Oxygen Concentration.

Phase	Days	Glucose-Loading (g m^{-2} d^{-1})	NH$_4$N-Loading (g m^{-2} d^{-1})	Oxygen (g m^{-3})
I	0–108	1.5	—	6
II	109–140	—	—	6
III	141–224	1	3	6

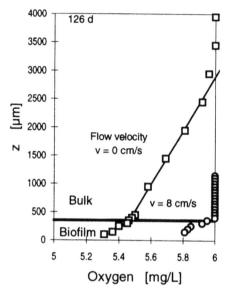

Figure 7 Oxygen profiles in a heterotrophic biofilm under different flow conditions.

($v = 0$ cm s^{-1}) could be used to calculate the oxygen flux into the biofilm according to Fick's first law. The oxygen flux was used to fit the kinetic parameters for the processes of maintenance and inactivation of the heterotrophic microorganisms in the biofilm model (Horn and Hempel, 1997b). This experiment demonstrated that the oxygen microelectrode could be used even when the concentration gradient lay below 0.02 mg per 100 μm distance under flow conditions. If the distance was great enough, the experimental data could be used to evaluate the kinetic parameters of the biofilm system.

The next aspect to be investigated was the transition from a heterotrophic biofilm to an auto-/heterotrophic biofilm in phase III. Again, the oxygen microelectrode could be used to work out the behavior of the microorganisms in the biofilm under defined substrate conditions. In Figure 8 the left side shows the measured and modeled oxygen profile in phase III on days 162, 191 and 222. On the right side of the figure, the modeled volume fractions of the inert material and the autotrophic and heterotrophic biomass are shown. The simulation was performed with AQUASIM (Reichert, 1994). The processes and the kinetic parameters for the simulation have been presented elsewhere (Horn and Hempel, 1997b).

During the measurement of the oxygen profiles, the biofilm was only supplied with ammonia substrate. By this means, the activity of the autotrophic biomass could be worked out. The transition from a heterotrophic biofilm to an auto-/heterotrophic biofilm could be shown to a very fine degree by the measured oxygen profiles. The oxygen profiles in the biofilm became steeper

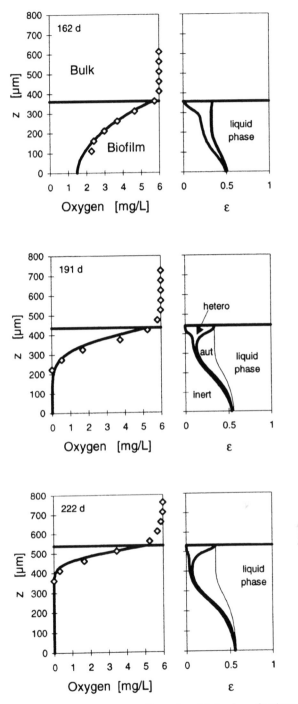

Figure 8 Measured and modeled oxygen profiles and modeled volume fractions of the considered particulate components in phase III (inert = inert material, hetero = heterotrophic biomass, aut = autotrophic biomass).

and steeper and the model simulated an increasing volume fraction of the autotrophic biomass.

The oxygen profiles are presented as mean values. The deviation of the oxygen profiles between the different ports of the biofilm tube reactor has been shown in previous studies (Horn and Hempel, 1995; Horn and Hempel, 1997a). Although the biofilm surface was, in fact, not flat and the structure was not as homogeneous as assumed in the biofilm model, the deviation of the measured oxygen profiles was in an acceptable range. This was due to the high density of the investigated biofilms. The mean biofilm density lay between 25,000 and 30,000 gm^{-3} biofilm volume. It could be shown that in biofilms that were grown faster, the density decreased and the deviation between the oxygen profiles decreased.

STRUCTURE OF HETEROTROPHIC BIOFILMS UNDER DIFFERENT SUBSTRATE CONDITIONS

The structure of biofilms could be investigated by confocal laser scanning microscopy (CSLM). By this method, the structure of the biofilm could be visualized in three dimensions. The question remained as to how structures of biofilms developed. Recent studies suggest that the substrate concentration is a main factor that determines the biofilm structure (Wimpenny and Colasanti, 1997).

Very simple experiments on the growth of biofilms support this claim. Figure 9 shows the dependence of the biofilm density on the substrate loading during the biofilm growth. These experiments need a certain time and could not be performed in only a few days. Substrate supply is not the only influ-

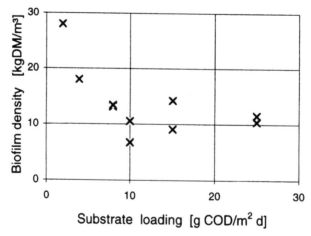

Figure 9 Dependence of the biofilm density on the substrate loading of heterotrophic biofilms.

ence on the biofilm density. With increasing biofilm age, the biofilm density also changes (Horn and Hempel, 1997a). Furthermore, the hydrodynamic conditions have to be considered when the biofilm structure is investigated.

The microelectrode could be used for purposes of studying the substrate conversion and the substrate transport in the growing biofilm. Figure 10 shows oxygen profiles in two heterotrophic biofilms that were grown with different substrate loads. Both biofilms were grown at a Reynolds number >4000. The biofilm density of biofilm (a) was 10,500 gm^{-3} and that of biofilm (b) was 15,000 gm^{-3}. The profiles were measured under oxygen-limiting conditions, to work out the different patterns of the profiles in the biofilms.

It could be seen that the profiles in the less dense biofilm (a) deviated from each other in the deeper parts of the biofilm. The penetration depth of the oxygen and, with it, of the substrate varied between 125 and 300 μm. The oxygen profiles of biofilm (b) were more homogenous, and the penetration depth differed only between 125 and 175 μm. This was due to the more compact structure that developed under substrate-limiting conditions over 6 weeks of growth. The less dense biofilm (a) developed in only 2 weeks.

The two oxygen profiles show the influence of substrate conditions on the development of the biofilm density and structure. Figure 9 can be taken as a first statement on biofilm development, but the influence of the hydrodynamic conditions has to be investigated further. The experiments conducted until now suggest that the biofilm density and the maximum biofilm thickness are a func-

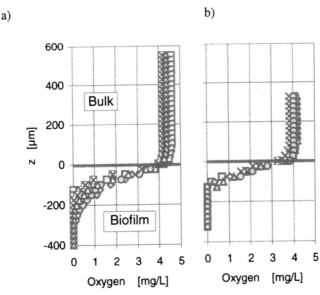

Figure 10 Oxygen profiles in two biofilms grown with different substrate load. Biofilm (a) 10 g COD m^{-2} d^{-1} and biofilm (b) 4 g COD m^{-2} d^{-1}.

tion of substrate concentration (C_S), oxygen concentration (C_{O2}) and hydrodynamic conditions:

$$
\left.\begin{array}{l}
\text{Biofilm density} \\[1em]
\text{Biofilm thickness}
\end{array}\right\} = f\,(C_S,\ C_{O2},\ \text{Re}) \tag{5}
$$

DISCUSSION

The Clark-type oxygen microelectrode has been a sensitive and useful tool in biofilm research. Several studies have been made in biofilms and in sediments. Combined with the CSLM, the microelectrode will remain the most important tool in biofilm research for some years. Nevertheless, the microelectrode technique has to be improved. New reactors have to be developed for in situ measurement of substrate and oxygen profiles in biofilms that have been grown under defined substrate and hydrodynamic conditions. The investigation of suspended biofilm pellets, in particular, is a great challenge. This means that more real biofilm systems from the technosphere and from the biosphere have to be investigated.

Apart from the discussion about the heterogeneity of biofilms, the available results from microelectrode studies show that, for several applications, it may be useful to treat the biofilm as a continuum. There is no question that the model concept of a flat and homogeneous biofilm surface has nothing to do with reality. Nevertheless, biofilm reactors have to be designed and substrate conversion and mass transport in environmental biofilm systems have to be modeled. Only in this way could the behavior of anthropogenic substances in the environment become better understood.

For example, the modeling of substrate conversion in biofilm reactors in wastewater treatment is of great interest. Changing substrate conditions in the effluent of wastewater treatment plants during the day and/or over the weekend, when the industrial production is lower, could be modeled with existing biofilm models. Figure 8 shows very impressively the transition from a heterotrophic to an auto-/heterotrophic biofilm. These results have improved biofilm modeling. Starvation of the microorganisms and changing substrate conditions are no longer a black box when modeling biofilms in wastewater treatment.

On the other hand, the modeling of biofilms in the environment has been developed on the basis of the presented laboratory results (Horn and Wulkow, 1996). In particular, the kinetic parameters for the autotrophic microorganisms could be used without adjustment. The mass transfer at the bulk/biofilm interface in small running waters could be described by adapting Equation (4) to the conditions in the stream bed.

The structure and, furthermore, the density of biofilms remain areas that are not well understood when biofilms in the technosphere and the biosphere are being modeled. It is in this area that biofilm research has to be focused. There is no doubt that the structure of biofilms is mainly influenced by the substrate conditions; this has to be considered when planning the setup for new experiments.

REFERENCES

Bungay, H.R., W.J. Whalen and W.M. Sanders. 1969. Microprobe techniques for determinig diffusivities and respiration rates in microbial slime systems. *Biotechnol. Bioeng.* 11, 765–772.

Cazelles, B., D. Fontvieille and N.P. Chau. 1991. Self-purification in a lotic ecosystem: a model of dissolved organic carbon and benthic microorganisms dynamics. *Ecol. Modell.* 58, 91–117.

Centonze, D., A. Guerrieri, C. Malitesta, F. Palmisano and P.G. Zambonin. 1992. Interference-free glucose sensor based on glucose-oxidase immobilized in an overoxidized non-conducting polypyrrole film. *Fresenius J. Analyt. Chem.* 342, 729–733.

Characklis, W.G. 1989. Laboratory biofilm reactors. In W.G. Characklis and K.C. Marshall (eds.): *Biofilms.* Wiley, New York, 55–89.

Cronenberg, C.C.H. and J.C. van den Heuvel. 1991. Determination of glucose diffusion coefficient in biofilms with microelectrodes. *Biosensors Bioelectronics* 6, 255–262.

Dalsgaard, T. and N.P. Revsbech. 1992. Regulating factors of denitrification in trickling filter biofilms as measured with the oxygen/nitrous oxide microsensor. *FEMS Microbiol. Ecol.* 101, 151–164.

Debus, O. 1993. Aerober Abbau von flüchtigen Abwasserinhaltsstoffen in Reaktoren mit membrangebundenem Biofilm. Aerobic degradation of volatile substances in membrane biofilm reactors. In I. Sekoulov (ed.): *Hamburger Berichte zur Siedlungswasserwirtschaft.* Hamburg: GFEU an der TUHH.

Fu, F.P., T.C. Yun-Chung, Zhang and P.L. Bishop. 1993. Determination of effective oxygen diffusivity in biofilms grown in a completely mixed biodrum reactor. *IAWQ: Second International Specialized Conference on Biofilm Reactors,* Paris, 543–550.

Glud, R.N., N.B. Ramsing and N.P. Revsbech. 1992. Photosynthesis and photosynthesis-coupled respiration in natural biofilms quantified with oxygen microsensors. *J. Phycol.* 28, 51–60.

Gust, G., K. Booij, W. Helder and B. Sundby. 1987. On the velocity sensitivity (stirring effect) of polarographic oxygen microelectrodes. *Netherland J. Sea Res.* 21, 255–263.

Hickey, C.W. 1988. River oxygen uptake and respiratory decay of sewage fungus biofilm. *Water Res.* 22, 1375–1380.

Hooijmans, C.M. 1990. Diffusion coupled with bioconversion in immobilized systems: use of an oxygen microsensor. PhD thesis, University of Delft.

Horn, H. 1992. Simultane Nitrifikation und Denitrifikation in einem hetero-/autotrophen Biofilm unter Berücksichtigung der Sauerstoffprofile. Simultaneous nitrification and denitrification in an auto-/heterotrophic biofilm with respect to the measured oxygen profiles. *gwf Wasser/Abwasser* 133, 287–292.

Horn, H. and D.C. Hempel. 1995. Mass transfer coefficients for an autotrophic and a heterotrophic biofilm system. *Water Sci. Technol.* 32, 199–204.

Horn, H. and D.C. Hempel. 1997a. Substrate utilization and mass transfer in an autotrophic biofilm system: experimental results and numerical simulation. *Biotechnol. Bioeng.* 53, 363–371.

Horn, H. and D.C. Hempel. 1997b. Growth and decay in biofilm systems. *Water Res.* 31, 2243–2252.

Horn, H. and M. Wulkow. 1996. Modellierung von Einleitungen in kleine Fließgewässer. Modeling of waste water discharge in small running waters. *gwf Wasser/Abwasser* 137, 557–564.

Kuenen, J. G., B.B. Jorgensen and N.P. Revsbech. 1986. Oxygen microprofiles of trickling filter biofilm. *Water Res.* 20, 1589–1598.

Kühl, M., Y. Cohen, T. Dalsgaard, B.B. Jörgensen and N.P. Revsbech. 1995. Microenvironment and photosynthesis of zooxanthellae in scleractinian corals studied with microsensors for O_2, pH and light. *Marine Ecol. Prog. Ser.* 117, 159–172.

Lens, P., M.-P. de Poorter, C.C. Cronenberg and W. Verstraete. 1995. Sulfate reducing and methane producing bacteria in aerobic wastewater treatment systems. *Water Res.* 29, 871–880.

Leu, H.-G., C.-F. Ouyang and J.-L. Su. 1996. Effects of flow velocity changes on nitrogen transport and conversion in an open channel flow. *Water Res.* 30, 2065–2071.

Lewandowski, Z., S.A. Altobelli and E. Fukushima. 1993. NMR and microelectrode studies of hydrodynamics and kinetics in biofilms. *Biotechnol. Prog.* 9, 40–45.

Lorenzen, J., R.N. Glud and N.P. Revsbech. 1995. Impact of miccrosensor-caused changes in diffusive boundary layer thickness on O_2 profiles and photosynthetic rates in benthic communities of microorganisms. *Marine Ecol. Prog. Ser.* 119, 237–241.

Reichert, P. 1994. *AQUASIM—computer program for simulation and data analysis of aquatic systems Bd. 7.* Dübendorf: Schriftenreihe der EAWAG.

Revsbech, N.P., L.P. Nielsen, P.B. Christensen and J. Sorensen. 1988. Combined oxygen and nitrous oxide microsensor for denitrification studies. *Appl. Environ. Microbiol.* 54, 2245–2249.

Sweerts, J.-P. and D. de Beer. 1989. Microelectrode measurements of nitrate gradients in the littoral and profundal sediments of a meso-eutrophic lake (Lake Vechten, Netherlands). *Appl. Environ. Microbiol.* 55, 754–757.

Wimpenny, J.W.T. and R. Colasanti. 1997. A unifying hypothesis for the structure of microbial biofilms based on cellular automaton models. *FEMS Microbiol. Ecol.* 22, 1–16.

Index

241